Feeding, pruning and pest control

Brian Walkden

Edited For U.S. Gardeners
By Marjorie Dietz

Floraprint

Published 1977 by Floraprint Limited,
Park Road, Calverton, Nottingham.
Designed and produced for Floraprint by
Intercontinental Book Productions
Copyright © 1977 Intercontinental Book Productions
and Floraprint Limited. North American edition Copyright © 1981 Intercontinental Book Productions and
Floraprint U.S.A.

ISBN 0-938804-04-9

Design by Design Practitioners Limited

Photographs supplied by Floraprint Limited (copyright
I.G.A.), Leslie Johns and Associates, Harry Smith

Cover photographs supplied by Floraprint Ltd.

Printed in U.S.A.

Contents

1 Basic gardening techniques   4

2 Feeding plants   6

3 Compost   16

4 Watering   22

5 Pruning   26

6 Weeds and weed control   41

7 Controlling pests and diseases   47

Index   64

1 Basic techniques

For every gardener, there are certain techniques that can be said to lay the foundations of a beautiful, healthy and successful garden – the techniques that, being unglamorous in themselves, are rarely the subject of a special book: feeding plants, using compost, watering, pruning and keeping gardens free of weeds, pests and diseases. Yet these are the very areas in which many gardeners feel their knowledge could be improved. Among the questions often raised are:

Why should it be necessary to feed plants?

What is the difference between the various types of compost?

How much should plants be watered?

How do I know where, when and how much to prune plants, and why does it have to be done?

How can I keep my garden weed-free without hours of back-breaking labour?

How do I recognize what pest or disease is destroying my plants, and what can I do about it?

The general principles for successful garden cultivation remain the same whatever

Compost is invaluable; site the compost heap in the vegetable plot.

the type of garden, but unless gardeners know the whys and wherefores, their own actions (for example, overfeeding, over-watering, or cutting back essential new growth) may be as much responsible for poor results as the ravages of weeds, pests and diseases. These points and the answers to the questions listed above are dealt with here in detail chapter by chapter.

2 Feeding plants

Successful plant growth depends on good feeding. Soils differ widely in their content of foods or chemicals that are beneficial to plants but there must be adequate organic matter (humus) in the form of the compost to keep the soil in good condition. In addition, sufficient artificial or chemical feeds will be needed to maintain a balanced growth.

Even in an uncultivated garden, soil is manured or provided with organic matter by the action of worms, which bring decomposed leaves into the subsoil. There is also a natural process of decay or rotting down within the surface vegetation. However, where the soil is required for intensive cultivation, it is necessary to apply plant foods in a more concentrated form and more frequently, since plants quickly use up the food that is in the soil.

Most soils contain reserves of essential chemicals that slowly become available to plants by the action of soil organisms and weathering. It is still necessary, however, to add to these chemicals. In many cases, more of one chemical than another must be added in order to provide the plants with a good 'diet', to encourage sturdy, healthy growth, high crop yields and beautiful floral displays.

Frequent plant feeding is required for most garden plants so that they receive chemicals that can be used straight away by their root systems or that will be available in suitable forms, after some basic changes, in as short a time as possible.

Plant foods

Certain elements are essential to good plant growth. At the top of the list are three elements – nitrogen, potassium, and phosphorus. There are also some that are known as trace elements. These include calcium, magnesium, sulfur, iron, manganese, zinc, copper, boron and molybdenum. The first three elements are the ones gardeners need to supply to the soil, as many of the trace elements are probably already there in adequate amounts and are, in any case, only required in quite small amounts in comparison with the essential ones.

All the prime foods must be applied as a *balanced* feed to plants. Too much of one or too little of another will cause poor growth. By feeding back into the soil the correct proportions of plant foods the balance is restored. The results can be quite amazing.

Nowadays, it is possible to check the soil quite easily to find out what foods are lacking and how much of one particular food is required to bring the soil back into a balanced state. This is done with a soil-testing

Even uncultivated gardens are enriched by the natural processes of fertilization. Earthworms work decayed plant matter into the subsoil, providing new organic material. Their tunneling also aerates the soil.

Pink-flowered hydrangeas are an indication of an alkaline soil, while blue-flowered hydrangeas indicate an acid soil.

kit, which costs only a few dollars and is a very good investment – especially when planting a new garden.

The test outfit provides a color code reference for small soil samples, which are treated with special chemicals in tiny test tubes. After shaking up the contents of each tube, the mix is allowed to settle and then a color comparison is made against a special chart.

The deficiency (if any) can immediately be read off, while a check-off column at the side of the chart tells how much of a particular fertilizer is required to bring the soil back to its correct balance. These tests are only relevant for vital plant foods – not for the so-called trace elements.

Acid and alkaline soils

When tackling the problem of plant foods it is also important to know whether the soil is acid, neutral or alkaline. This condition will have an influence on the fertility of the soil and on what types of plant can be cultivated.

Plants such as the heathers and rhododendrons will grow well only in soils that are relatively acid. Cabbages and Brussels sprouts, on the other hand, will grow successfully only in soils that are neutral, slightly alkaline, or slightly acid.

Fortunately, there is a very simple way in which the acidity of soils is classified. This is by a scale that measures the soil's pH value. A reading of pH $7 \cdot 0$ indicates that the soil is neutral – neither acid nor alkaline. A reading above this figure indicates that the soil is alkaline, while below $7 \cdot 0$ the soil is acid.

The best values for most garden plants lie between pH $6 \cdot 5$ and $7 \cdot 0$. To make the soil more alkaline, lime must be added; to make it more acid, flowers of sulfur are applied. An example of the effect of an acid or alkaline soil is demonstrated in the case of colored hydrangeas. If blue flowers are produced all the time, this is an indication that the soil is acid. If, on the other hand, the flowers are pink, this is a sure sign that the soil is alkaline.

The degree of acidity or alkalinity of soils can be determined with a soil-test kit. A pinch of soil is mixed with the test chemicals and the resulting color matched against a standard chart. Each color indicates the degree to which a soil is acid or alkaline.

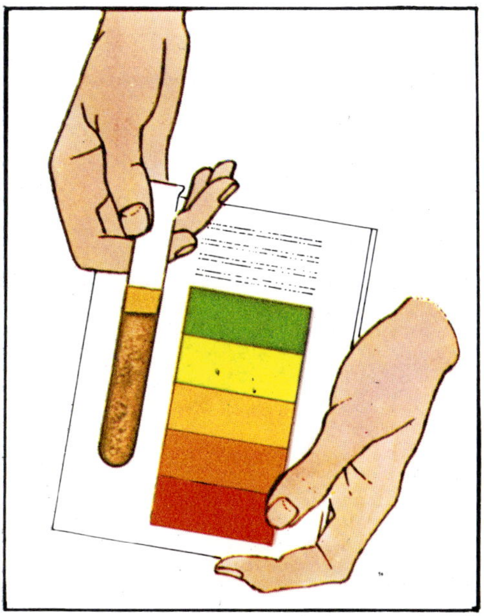

Let us now take a closer look at the individual plant foods that can be used to good
effect in the garden. The nitrogenous feeds
are generally very quick-acting and the
effect on growth can be appreciated soon
after application. The rate of growth
increases, leaves take on a dark green color
and become large and luscious (fleshy). By
comparison, lack of sufficient nitrogen
results in small, stunted plants and poor-
sized foliage, which may also take on a pale
bluish color.

Phosphorus is needed to promote good root development, without which the plants cannot obtain enough good food from the soil. It is a very important substance for root vegetables too. The phosphorus content in the soil is vital for seedlings to form a vigorous root system – thus building up an excellent plant for planting out later on in permanent quarters. Phosphorus is also essential for the satisfactory ripening of fruits, as is potash – an even more important plant food in this respect.

Potash adds color to fruits, flowers and certain vegetables. It also encourages better ripening and enhances the quality of food crops in terms of flavor and storing properties. This is especially the case with apples, pears, potatoes and carrots. Plants deficient in potash become poor bearers and yield smaller crops. Leaf margins – especially those of fruit trees – can become scorched where potash is lacking.

The other top-priority food for plants is calcium, which already exists in many soils in adequate quantities. Calcium helps to stimulate essential bacterial action in the soil, which makes other foods readily available to the plants. It also helps to improve the physical condition of soils, corrects acidity and provides a good 'grounding', as it were, for other plant foods.

Fertilizer should be scattered along rows and should be raked in well.

Applying foods

Plant foods can be applied to the soil as individual types, or as balanced or specially blended mixtures that contain all the essentials, such as nitrogen, phosphorus and potash, and also contain several important trace elements. Nitrogen is available as a plant food in the form of such fertilizers as sulfate of ammonia, nitrate of soda and urea. Phosphorus can be obtained as superphosphate and as bonemeal, which should be of the steamed kind. Potash is supplied in the form of sulfate of potash. This is the best form of potash for the gardener as it is also relatively quick-acting.

Here is a checklist of some of the most popular plant foods. Others exist that may only be available locally. For instance, for those who live near the sea, gathering fresh seaweeds (a source of nitrogen and potash as well as humus) is perfectly feasible. Other fertilizers, such as cottonseed meal, which is popular with organic gardeners who avoid the commercial fertilizers, are often used in general purpose organic fertilizers as a source of nitrogen. Organic fertilizers are slow to break down, but usually less drastic than artificials.

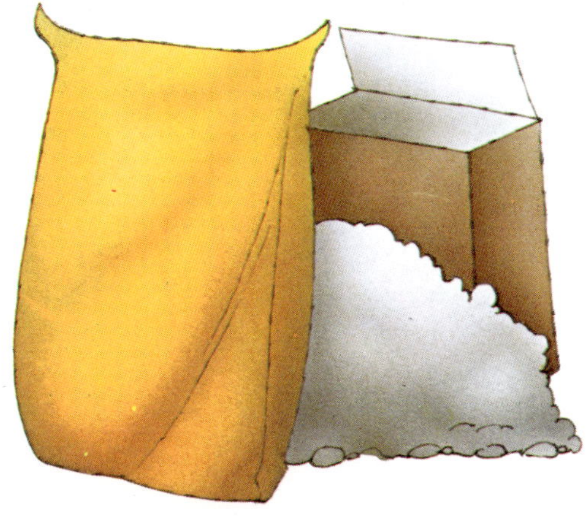
Foods can be purchased individually or they can be bought in pre-mixed bags.

Popular plant foods

Animal manures These include poultry, cow, horse, goat and others. They were once the main source of fertilizers and are still important in rural regions where they are most likely to be available. Cow manure is also usually available in dehydrated form from garden centers. Most manures, especially horse and cow, are valuable sources of humus as well as being general purpose fertilizers, especially when they are mixed with litter – straw, peat moss. Fresh manures are safe to incorporate with the soil in fall. Decayed manures are safest in spring.

Bonemeal Provides phosphorus (20–30 per cent) very slowly and a very small amount of nitrogen (1–4 per cent) fairly quickly. Steamed bonemeal is fastest acting. Apply at rate of 4–8 oz per sq yd (110–220 g per sq m).

Cottonseed meal Contains about 7 per cent nitrogen, 3 per cent phosphorus and 2 per cent potash. A long-lasting source of nitrogen, especially recommended for rhododendrons and its relatives. Apply 6 lb per 100 sq ft (2·7 kg per 9·3 sq m).

Dried blood May contain 12 per cent nitrogen. Fairly fast-acting but it is expensive. Save for salad crops. Use at rate of 2–3 oz per sq yd (56–84 g per sq m).

Limestone Not basically a fertilizer, although it is a source of calcium, it is used to counteract acidity. Limestone is slower acting than hydrated lime. Gypsum (calcium sulfate) is applied to soils to supply calcium but not alter the acidity. Gypsum is excellent for the physical improvement of soils, especially heavy clay types, as it helps to break them down.

Nitrate of soda This is a highly nitrogenous fertilizer (about 16 per cent nitrogen) that is quick-acting. It can be used very successfully during major growing periods and is ideal for leafy plants such as lettuce and spinach. Care must be taken when using it as it can burn plants, especially young ones. Rate of application is about ½–1 oz (14–28 g per sq m). It is perhaps easiest to apply diluted in water at the rate of ¼–½ oz per gal. (7–14 g per 4 liters) of water.

Sludge This is the dried and pulverized residue from sewage disposal plants and may well end up the fertilizer of the future. The best known sludge that has been marketed with any success is Milorganite, a good all-purpose fertilizer high in nitrogen. Follow instructions on the bag.

Sulfate of ammonia Similar in use to nitrate of soda. This plant fertilizer is for boosting green growth. It is not so caustic as nitrate of soda and is best applied in spring and early summer. Can be mixed with sulfate of potash and superphosphate. It contains about 20 per cent of nitrogen. Apply at the rate of ½–1 oz per sq yd (14–28 g per sq m).

Sulfate of potash This contains about 50 per cent of potash and acts quite rapidly and is noncaustic. Can be used at any time of the growing season. Apply at the rate of ½–1 oz per sq yd (14–28 g per sq m). Muriate of potash, containing about 60 per cent potash, is similar.

Superphosphate For supplying phosphates to plants in the quickest form, this is the fertilizer to use. It is less costly than bonemeal. It should be cultivated into the soil at planting time at rate of 1–3 oz per sq yd (28–84 g per sq m). The amounts of phosphoric acid vary from 16, 20, 32 to 44 per cents.

Wood ashes A source of potash. The percentage varies but about 5 per cent of potash is typical. Use in spring or fall at rate of about 4–6 lb per 100 sq ft (1·8–2·7 kg per 9·3 sq m). Wood ashes can burn when first applied, so do not put on plants in active growth.

Mixing your own plant foods

It is possible to take several different fertilizers and combine them to form an excellent balanced plant food.

While the practice is widespread in Europe and in U.S. commercial growing operations, it does not have a great deal of practicality for the hobby gardener.

First of all, locating sources of these various chemicals can prove to be troublesome. Secondly, there is a risk of not mixing these accurately, so that either the plant is burned by excessive amounts applied or the dosage is too small to provide any noticeable benefit.

If, however, you like to experiment, there are several vegetable and fruit crop formulas shown on this page that you will want to try.

Roses. Superphosphate, 12 parts by weight. Sulfate of magnesium, 2 parts by weight. Nitrate of potash, 10 parts by weight. Sulfate of iron, 1 part by weight. Gypsum, 8 parts by weight. Mix well and apply in April at the rate of 4 oz. per square yard.

Tomatoes. Superphosphate, 3 parts by weight. Sulfate of ammonia, 2 parts by weight. Sulfate of potash, 2 parts by weight. This mix is ideal as a feed, while the plants are growing. As soon as the first fruits have set and begun to swell, give each plant one teaspoonful around its base and repeat every 10-14 days. Water in well afterwards.

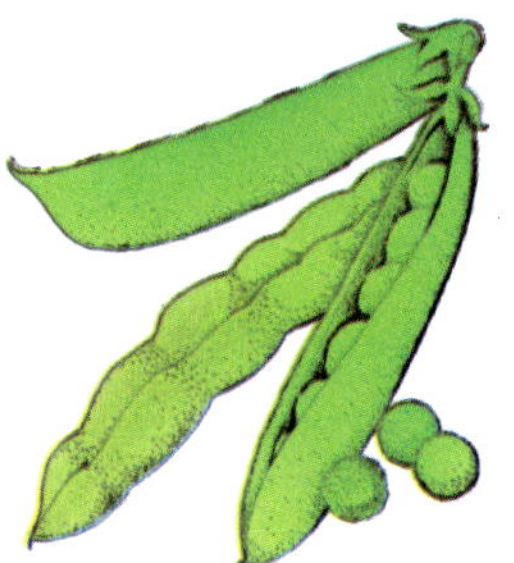

Peas and beans. Sulfate of potash, 2 parts by weight. Sulfate of ammonia, 1 part by weight. Superphosphate, 3 parts by weight. Use just before seeds are sown, at 2 oz. per square yard.

Cabbages, Brussels sprouts, broccoli. Superphosphate, 3 parts by weight. Sulfate of ammonia, 2 parts by weight. Sulfate of potash, 1 part by weight. Mix well and use before sowing at the rate of 3 oz. per square yard.

Fruit trees. Sulfate of potash, 1 part by weight. Sulfate of ammonia, 2 parts by weight. Apply at the rate of 3 oz. per square yard in early spring.

Root crops (Carrots, beets, parsnips). Superphosphate, 4 parts by weight. Sulfate of ammonia, 1 part by weight. Sulfate of potash, 2 parts by weight. Mix together thoroughly and use before seed is sown at the rate of 4 oz. per square yard.

Rates of application are given so you can make up as much of each formula as required to treat a given area. Amounts of each chemical to be used in the various formulas are given by weight. For example, if 1 lb. is the basic weight measure to be used, a formula calling for 5 parts by weight of a given chemical would become 5 lb.

A general garden formula and a specialty lawn food for poorly growing grass are also given on the next page.

Plant-food formula for general garden applications

This is a well-balanced formula for all types of garden plants and is especially useful in flower borders.

Superphosphate, 7 parts by weight; sulfate of potash, 2 parts by weight; sulfate of ammonia, 5 parts by weight; and bonemeal, 1 part by weight. Mix ingredients thoroughly and apply at the rate of 3-5 oz. per square yard. It can be used as a general plantfood or during the preparation of soil before sowing takes place.

Lawn Food

This formula is especially recommended for lawns in poor health: Sulfate of potash, 2 parts by weight; sulfate of ammonia, 1 part by weight; dried blood (available commercially as a meal), 2 parts by weight; and sharp sand, 20 parts by weight (this helps add bulk to the mix and also makes it easier to distribute). Apply at the rate of 5-6 oz. per square yard in the spring and, if possible, add a further dressing about 5 weeks later. Apply only when the ground is moist or just prior to a rainfall.

Commercial fertilizers

There are a number of commercial fertilizers that are available to the gardener. These include dry and liquid types that can be spread or sprayed on a variety of crops and ornamental plants for a number of different purposes.

Most are convenient to use and because the formulas have been widely tested, crop response will be as expected, unless the gardener does not follow instructions or some chemical imbalance exists in the soil.

Many fertilizers incorporate weed killers. These "weed and feed" formulations are usually in the form of granular or pelleted dry compounds and are evenly applied with the aid of gravity flow or broadcast spreaders.

Liquid fertilizers can also be applied with herbicides mixed in the solution. A proportioner jar attached to a hose does a good job of applying these solutions.

Formulas exist that regulate the amount of plant food made available at one time. These fertilizers have a time release feature that spreads the availability of plant food up to a year. Ornamental trees and shrubs benefit greatly from such fertilizers. On the other hand, flowers and vegetable crops, growing rapidly in the garden or containers, demand frequent feedings. Although fertilizers geared to these crops provide a relatively small amount of plant food, nearly all of it is available immediately.

There are also starter fertilizers in commercial potting soils, helping to germinate seed and producing strong seedlings. As plants grow, however, a continuous supply of plant food must be provided. Not only will your plants perform better, but resistance to disease will be increased.

Already mixed dry fertilizers can be spread evenly onto lawns using gravity flow or broadcast spreaders.

Special plant foods

A great deal of progress has been made in recent years in the preparation of other types of plant foods. There are some based on specially treated seaweed. Several of these are liquid types which are diluted in water according to instructions. A large range of trace elements is a feature of seaweed plant foods and excellent results can be obtained using them.

Other plant foods include the processed natural manures originating from chicken, cattle and sheep feeding operations. They are dried and pulverized to form an easily applied product which, in most cases, does not smell badly either. Dried manures provide a good source of plant food and yet are relatively economical to use. These products have the additional benefit of containing a useful amount of organic matter, which helps to condition the soil as well as encouraging the formation of a vigorous root system.

Plant foods should be applied at regular intervals during a plant's development. The aim of this action is to promote sturdy growth and to encourage an abundance of fruits or flowers. Feeding generally starts as soon as the plants are well-established and, in the case of flowers, when the buds are forming. This type of feeding differs from the incorporation of fertilizer when the bed is being prepared to receive a crop.

Many gardeners find that the various liquid plant foods are valuable for maintaining plant health and encouraging production of bountiful crops of flowers or vegetables. Often, liquids are applied directly to foliage. The diluted material is watered on by a can having a fine spray nozzle or by a spray bottle attached to a hose.

However, plants will benefit from liquid

Plants will absorb nutrients more rapidly if fertilizers are applied in a dilute liquid form by watering can.

Another method is to spray the liquid feed over the leaves with a syringe. This is known as a foliar feeding.

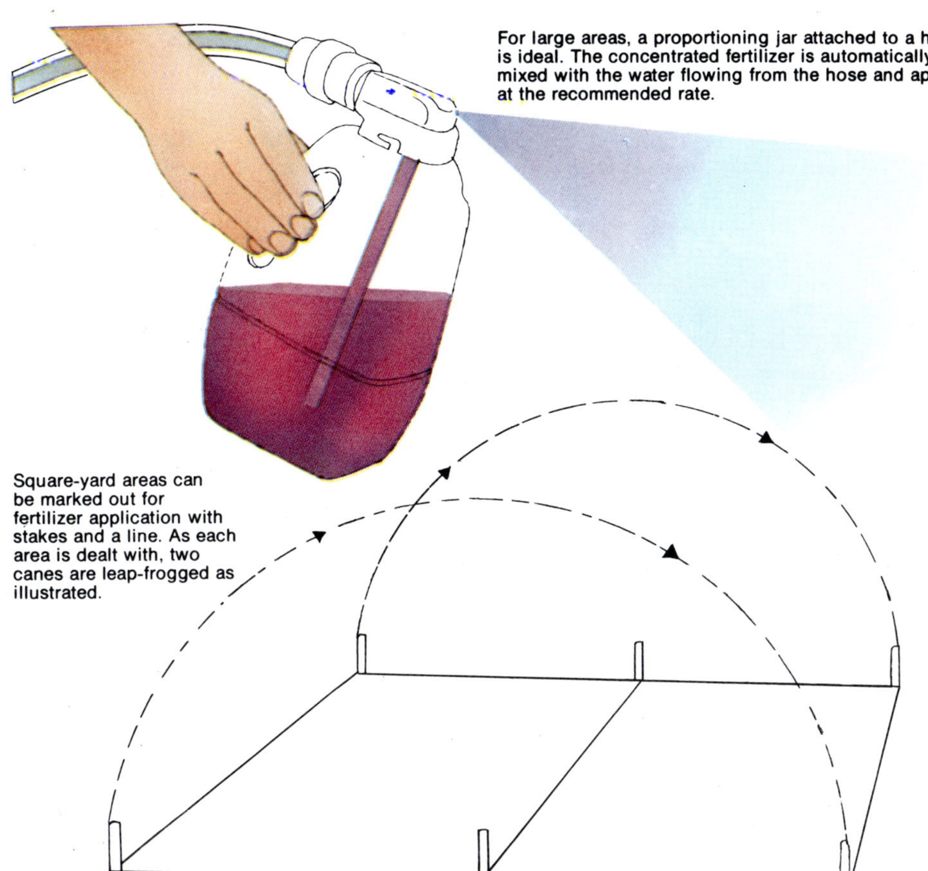

Square-yard areas can be marked out for fertilizer application with stakes and a line. As each area is dealt with, two canes are leap-frogged as illustrated.

fertilizers applied to the soil. This method of application is particularly desirable to gardeners who do not wish residue from foliar applications remaining on leaf surfaces. Yet the plant food is in a form that is easily taken up by the roots and a similar 'quick response' is noted, as when foliar feedings are made.

Accurate and even application of fertilizers is very important. A simple way to make certain application is being done accurately is to mark out the ground in square yard strips. Do this by marking out two parallel lines with garden twine, spaced 1 yard apart. Then place two stakes across these two lines, spacing them 12 in. apart. Mark out the area as a square yard and scatter fertilizer evenly over it at the prescribed rate.

Of course, measurements can be changed to suit each situation. As an area is treated, one stake is brought forward to mark the next section. Work proceeds in this 'leap-frog' fashion until the entire area has been treated.

Although this method is impractical for fertilizing most lawns, it is useful when renovating relatively small areas, and in the garden, prior to planting.

Broadcast and gravity flow spreaders are most efficient for spreading fertilizer accurately over a large area. The most thorough, even coverage is achieved by walking the lawn in two different directions, but with the spreader set at only half the recommended rate.

3 Compost

In gardening terms, there are two types of compost. The first is the special soil mixture that is made up for the cultivation of a wide range of plants. Then there is the compost we are concerned with here – the finished product, as it were, that results when waste vegetation is collected and allowed to rot. Both kinds are invaluable for soil and crop improvement.

Quite often compost is referred to as humus material. It is, without any doubt, a most important product and is essential if good results are to be expected from your garden, no matter what you are growing. It is something that is easy to provide because most of the ingredients for making compost

A garden can virtually feed itself. All the essential materials for making good compost can be obtained from the plants raised there. Trimmings from the vegetable plot can be used for compost as can leftover kitchen scraps.

Bracken, seaweed, leaves, dead flower heads and other trimmings can all go into the compost heap. Such material is a rich source of chemical nutrients.

actually come from the garden itself.

Any soft material can be used to form compost. This includes hedge trimmings, grass cuttings, trimmings from the vegetable garden, autumn leaves and many waste scraps from the kitchen – for example, tea leaves, peelings, etc. When the waste material is decomposed efficiently it becomes a dark brown, fairly friable mass.

Of course, if it is not made correctly, compost can become a rather nasty evil-smelling pile of slimy material that is pretty useless. In these days of natural manure shortages it is vital that as much compost as possible be made for the garden. Here is how to go about preparing your own.

Making compost

The amount of compost you will be able to make depends on the quantities of basic material you can obtain from your garden. It is surprising how many grass mowings can be collected over the cutting season. This material can form quite a large part of a compost-making scheme. If you have a busy vegetable plot there will be lots of 'bits and pieces' you can gather here, such as leaf trimmings from cabbages, lettuces and Brussels sprouts, and the tops of various root crops such as carrots, turnips and beets.

Then there are the bean and pea growths (haulms, as they are technically called) that can be added to the compost when the crop has been gathered. When you tidy up the garden, cutting off the dead flower heads, for instance, you can add the refuse material to your compost. In the autumn, there is usually a great clearing up effort made when the dead top-growth of plants, especially in the herbaceous borders, is removed. All this can be used to make compost.

Most weeds can also be incorporated. Try to place weeds in the compost before they have made their seeds otherwise you will be sowing more weeds in the compost, and once such compost has been dug in, a

It is amazing how much rich compost material can be obtained from the clippings gathered in a mower's collecting box.

large proportion of the weed seeds it contains will grow again.

It is possible that you may be able to obtain other materials for your compost-making. Some bracken and seaweeds are useful but are best used in a moist condition, as they will not rot down easily if too dry. Small amounts of straw can be included too. This should be wet or watered before it is incorporated. Thoroughly made compost should not contain many viable seeds.

There are a few items that should not be used in making compost. These include tough stumps or roots, and bits of plastic. Very hard prunings or trimmings from hedges are not suitable either, as they do not decompose well. Nor should lawn cuttings from grass recently treated with a weedkiller be used.

Material for compost-making should be placed in a compost area or heap in order to rot down. The success of a compost pile depends to a great extent on the way in which the heap is made and contained. The use of special compost bins or containers is thoroughly recommended. These not only hold your compost heap neatly – this is very important in smaller gardens – they also encourage more efficient decomposition.

The autumn fall of dead leaves can also be used in the compost heap. Hedge clippings, but not the woody twigs, are another valuable source of compost matter.

A simple compost bin can be made from wire netting attached to stakes.

Compost containers The material in a compost heap is broken down by bacteria and many other organisms that will only work well if there is adequate air, moisture, nitrogen, non-acid conditions and plenty of warmth. The modern compost bin is specially designed to promote the best conditions for the activity of bacteria. Holes in the sides of the container ensure adequate air penetration.

Of course, you can always make your own enclosure and this is what most gardeners do, using a portion of snow fencing, chicken wire or boards. The simplest container consists of wire or plastic mesh netting supported at four corners by strong posts driven into the ground. The smallest practical and efficient size is a little over 3 ft × 3 ft (1 m × 1 m). Usually a heap that is some 7 ft (2 m) in length and about the same in depth is the best size.

A compost heap should have a base made of coarse matter that permits proper drainage.

The first layer of material about 9 in (23 cm) deep is then laid in the bin.

The next layer is made up of manure or a rich fertilizer that serves to accelerate the decaying process.

The 'sandwich' effect is achieved by adding additional layers of plant matter and accelerator.

Avoid placing your compost heap in a low-lying damp location. Good drainage is important. Avoid deep shade or a place where there is no free movement of air, for here stagnant conditions may prevail. On the other hand, a place directly in the sun will cause the material to dry up too quickly and poor rotting will result.

Building the heap No matter what type of container is used, always start the heap with the coarsest or roughest materials so that they form a natural drainage system for the rest. This material will rot eventually but initially will do its work well. After this your heap is built up 'sandwich fashion' – that is to say, a layer of waste material followed by an application of a special compost accelerator and so on.

The accelerator is a special preparation that encourages a more rapid decomposition of the waste and ensures a better and sweet-smelling end product. The depth of

each layer should be about 9 in (20 cm).

Place the material lightly on the heap – ideally, scatter it over the surface with a fork. Also, try to balance the material so that the weight distribution is fairly even. This will prevent heavier areas from settling down more than lightly loaded ones. Also, decomposition will be encouraged if a balanced heap is made up.

The other big advantage with the use of special accelerators is that there is no need to turn the compost heap to ensure even and thorough decomposition.

Natural forms of accelerator include fresh manure – chicken, horse or farmyard manure. This can be applied on each 9 in

The ingredients of a compost heap must not be allowed to dry out. Water the heap when it becomes necessary.

(20 cm) layer of material. Such a heap will need turning to ensure thorough rotting down. This can be done after some six to eight weeks. Throw the material to one side, making sure that the outer sections are placed towards the middle of the new heap to rot down better. Keep the heap tidy and compact by retaining it in a wire enclosure – the original one can be used.

You can make up your own compost accelerator by applying ½ oz per sq yd (14 g per sq m) of sulfate of ammonia to the surface, then covering it over with about 1½ in (4 cm) of soil. This should be done after each 9 in (20 cm) deep layer of material has been laid down on the heap.

Rotting down will be much quicker in the warmer summer and early autumn months than in winter and early spring. A heap started in early spring will be ready to use

Quicker decomposition in the heap is obtained by scattering an accelerator in with the plant material.

some time in the summer. With a good accelerator, many heaps of soft material will be well rotted after seven weeks. A heap started in summer should be ready in the late autumn. For spring use, a winter heap can be made up.

Keep an eye on the layers as they are built up. If they look rather dry, give them a light watering, using a fine nozzle on the hose or on the spout of a watering can.

If the weather is exceptionally wet, it is a

This specially designed container has removable sides.

Both light and heavy soils will benefit by digging in compost. This helps to retain moisture and aerates the soil.

heap has rotted down well, the bin can be removed and a new heap started.

The benefits of compost

Prepared compost will do many things for your garden. In the first place it is a soil conditioner. Light soils tend to dry out quickly, especially in the summer. If plenty of compost is dug in when the ground is being prepared in the winter or spring, and at any time when beds are being made ready for sowing or planting, the compost will act like a sponge and retain valuable moisture for the plants' roots.

Heavy clay soils are those that have their soil particles packed tightly together. If compost is worked in frequently it will help to space out these particles and make the soil easier to work or cultivate. It takes a little time for an improvement to be noticed but it is well worth doing.

Roots Compost also encourages the formation of vigorous roots, which in turn produce a healthy plant, one that is capable of taking in more food and water. Lining the bottom of planting holes or drills with compost is an excellent way to ensure a good start for plants. The larger holes for trees and shrubs can be treated likewise.

good plan to provide a temporary roof over a heap in the form of a sheet of plastic or sheet of corrugated iron to keep off excess water. Some commercial bins or containers have a sort of top or roof. Some also have removable sides so that the progress of the heap can be judged and also so that the contents can be easily removed. When one

Rotted compost matter should be incorporated into the bottom of the trenches made while preparing a planting site, as it provides a source of nutrients.

Compost is ideal as a mulch or dressing along plant rows. It keeps the base of plants and their roots moist.

Quick-growing crops such as lettuce depend on a high moisture content in the soil. The inclusion of compost in the bed preparations for these crops will promote rapid growth, which in turn will result in crisp leaves.

Top dressing Compost is ideal as a mulch or top-dressing around plants or along plant rows. Applied to a depth of about 2 in (5 cm), the covering will reduce water loss from the surface of the soil and also suppress a lot of tiny weed seedlings. For many vegetable crops, mulching with well-rotted compost is an excellent idea. Crops that benefit especially from this system are peas and beans.

Root crops, such as carrots, benefit from a generous layer of compost in their drills at seed-sowing time. The compost retains soil moisture and encourages the formation of first-class tender roots.

Lawns Composted vegetable waste can be used for the preparation of lawn sites if the ground is enriched with the material by forking it in as the soil is cultivated. In light, sandy types of soil, this preparation is vital if the grass roots are to grow strongly and are not to suffer from very bad dehydration during hot weather.

Under glass Where plants are grown under glass – in greenhouses and cold frames – enriching of the soil beds is important, as they tend to dry out more quickly under glass owing to the persistent high temperatures. For pots or other containers that are used for tomato and cucumber cultivation, the compost can be mixed with soil at the rate of one part compost to three parts soil. This, and some basic fertilizers, will produce an excellent growing medium with which to fill the containers.

When preparing soil borders or beds, outdoors or under glass, the same proportion can be used. Wherever the soil is light and tends to dry out badly the compost ingredient can be increased by one part.

This also applies to the use of window boxes and larger containers or tubs for flower displays. The one-to-three mix has proved successful for a wide range of plants when used with fertilizers.

Compost is also useful as a top-dressing for lawns, where it is applied in the autumn or spring. The material is scattered as evenly as possible over the surface of the grass at the rate of a small to medium-sized barrow load to every 3–4 sq yd (3–4 sq m). Afterwards, the compost is carefully worked or 'rubbed' into the surface with to-and-fro motions using the back of a rake or a stiff brush. A large proportion of the compost will be incorporated in this way and the remainder will gradually be washed in by the rain.

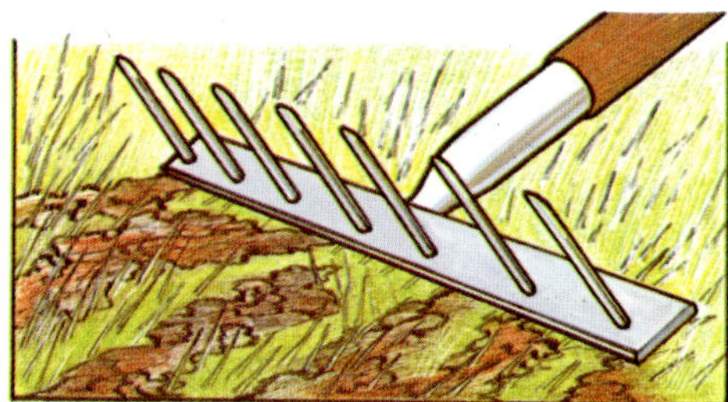

The lawn will benefit from a dressing of compost worked into its surface, especially in sandy soils that tend to dry out rapidly and that tend to be relatively poor in nutrients.

4 Watering

One of the most important gardening operations is watering. A plant must have a regular intake of water in order to grow well. In the case of fruit-bearing crops water is essential in order to swell the fruits.

Watering is vital where plants are being grown under glass and where they cannot benefit from natural rainfall. Plants grown in containers under glass dry out surprisingly rapidly, especially in high temperatures. Without the correct amount of water all plants will suffer, since the nourishment that the plants are obtaining from the soil is in solution. Even cacti and succulents need plenty of water at certain times.

The art of watering Watering is something of an art, because too little will hinder good growth whereas too much can cause the soil to become waterlogged so that the plants' roots do not grow well. Too much water in the ground also cools the temperature and chills the soil, making for very poor plant progress.

If there is any 'golden rule' it is that the gardener should water little and often: the aim should be to keep the soil just nicely moist at all times. Remember that fruit-bearing plants such as tomatoes, cucumbers and melons need increasing amounts of water as their fruits swell.

Special attention to watering is required for crops under glass (left) and especially such plants as melons, tomatoes and cucumbers. Where adequately watered, rapid swelling of the fruit will result.

In the greenhouse a good test for watering requirements is to pick up a pot and give it a tap. If there is a ringing note the pot needs watering. A dull ring, on the other hand, means that the soil is wet enough. This system applies only to the old (clay) type of pot. Unfortunately plastic ones provide no such practical clues.

However, a well-watered pot always feels heavier than a dry one, so this could be a fair means of gauging the condition of plants in plastic pots. It is now possible to purchase a special moisture indicator from your local garden shop or garden center. The probe of this device is pushed into the soil in the pot, and straightaway it will indicate whether conditions are wet, moist or dry. The device works from tiny batteries that will last many months. This is a very handy gadget, especially for house plants, which need constant checking for their watering requirements.

In the home the dry atmosphere of a

A moisture indicator (below) will reveal the moisture level of the soil at a glance.

In the warmth of the home, stand pot plants (above) on top of a water-soaked gravel bed in order to give them more moisture.

heated room quickly dries out the soil in plant containers. One method of overcoming this is to place plant containers in trays or dishes filled with small pebbles or gravel. Run in some water so that the level is just below the stones. If the pot is carefully placed on the stones its base will not rest in the water causing the soil to become water-logged. However, the air around the container will be kept cool and moist and will provide a better growing atmosphere for the plant.

Methods of applying water

The two most common methods of watering are by hose and by watering-can. For indoor use a smaller capacity can with a long spout is much easier to handle. Some gardeners use their thumbs to control the force of water from a hose, but it is better to have a nozzle that can easily be turned to provide a spray or jet of water. It should never be so forceful that it washes soil from the roots.

With special connectors, a permanent hose system can be laid out for watering large gardens. Watering points can be attached so that all parts of the garden are watered at once.

To conserve soil moisture, black polyethylene sheeting can be used as a mulch. Slits are cut for the plants to come through.

Sprinklers For watering large areas, such as a lawn or vegetable garden, sprinklers are a necessity. These range from the very simple stick-in-the-ground designs with a central hole through which the water is emitted in a spray pattern, to the more sophisticated adjustable sprinklers that move their spraying bar from side to side and whose direction is controlled by a switch system. The latter can be 'dialed' or set to water certain patterns. Their greatest advantage is that application can be controlled with accuracy.

Various kinds of irrigation systems using freeze-proof plastic pipes for above or below ground exist. Quite a complex layout can be quickly set up to reach various watering points in the garden. There, a tap can be attached so that only short lengths of hose are required to water any part of the garden.

Watering requirements can be reduced if the soil around plants is mulched. Rotted manure or waste vegetation from the compost heap can be used for mulching, applying it over beds or around plants to a depth of at least 2 in (5 cm). This prevents a lot of water evaporation from the surface of the soil. Black polyethylene sheets are also used for the same purpose. Turning a little soil over the edges as the sheets are laid down will hold them in place.

Watering under glass In the greenhouse, in gardens, and in frames automatic watering devices can play a useful role, especially during vacations. A simple system is one known as capillary matting. Capillary mats are capable of absorbing and retaining a considerable amount of water. If laid on the benching or staging with pot plants standing on top, each plant will be able to draw up its own water requirement.

The mat is kept supplied with water via a tank fitted with a simple float device. This piece of equipment is attached to the water trough, into which one end of the mat is placed. The mat absorbs water from the trough. To be completely automatic, the water tank can be attached to the main water supply. The floating ball valve in the tank will always keep it topped up with water.

For another simple system, a perforated hose or small-bore hose is used, fitted with an adjustable nozzle that can be regulated to provide a drip of water to plants in pots or in soil borders. This device can also be connected to a water-supply tank.

In a frame or among vegetables a trickle irrigation line with adjustable nozzles can be laid out and connected to a hose. The tap can be turned to slightly open to maintain a trickle of water to the system. Shading glass, whether in a greenhouse, or frame will reduce the amount of evaporation from the soil.

Watering times

Water should not be applied during the heat of the day. It is far better to water in the evenings or early in the morning before the sun is high. Make sure, too, that the outdoor soil is *thoroughly* watered. It may look wet and dark, but it may well be quite dry below the surface if you scratch down a bit.

Newly-bedded plants should be well watered, especially larger plants such as trees and shrubs. A good idea is to presoak the planting hole a few hours before placing the tree or shrub in position. As light soils tend to dry out more rapidly than others, they in particular *must* receive plenty of water.

In the open garden it is possible to apply a liquid feed at the same time as watering. This is accomplished by using a simple diluter attached to the end of the hose. The concentrated feed is put into the diluter and as the main flow of water passes through the hose the correctly diluted liquid is emitted from the end as a fan-like spray.

Capillary matting and a self-regulating water-supply device will provide an automatic watering system that is ideal for the greenhouse.

5 *Pruning*

Why do we need to prune? Basically, there are three very good reasons. The first is to produce the best possible fruits or flowers. The second is to maintain a tree or bush in good shape, and the third is to keep the plant as healthy as possible by removing dead or diseased growth.

There are, of course, several other advantages with pruning. It will keep a tree or shrub within predetermined bounds and will also maintain the attractive shapes that are so essential for fruit trees grown as cordons or fans.

The simplest forms of pruning are carried out purely to keep a tree or shrub in good shape. However, when fruit trees are pruned it is essential that certain rules be followed, as each type has its own special requirements. If these are not followed closely, a loss of fruit may result, simply because the branches that would have borne that fruit have been cut off.

Pruning is used to train such trees as this cordon apple. Unwanted shoots are pruned back to maintain the desired shape.

Raspberry canes that have fruited (dark outlines) should be cut right back to soil level. In general, cut canes back after harvesting. Cut everbearers that fruit in summer, autumn and again the next spring, only after the second crop has been harvested.

Pruning tools Successful pruning also depends on using the correct tools. These must be of good quality with keen cutting edges and an easy action so that they do not 'tire' when used for long periods. A poor cutting edge will also produce a ragged cut, liable to attract disease and infection.

There are two types of pruning shears. One has a so-called 'anvil cut'. Here the pruner has a sharp upper blade that comes down onto a bottom flat-edged blade. Pruners of this kind that have a slight sliding action are particularly good, as they produce a natural, knife-like cut. The other type of pruner cuts with a scissors action. The two blades cross over each other as the cut is made. Both upper and lower blades have sharp cutting edges.

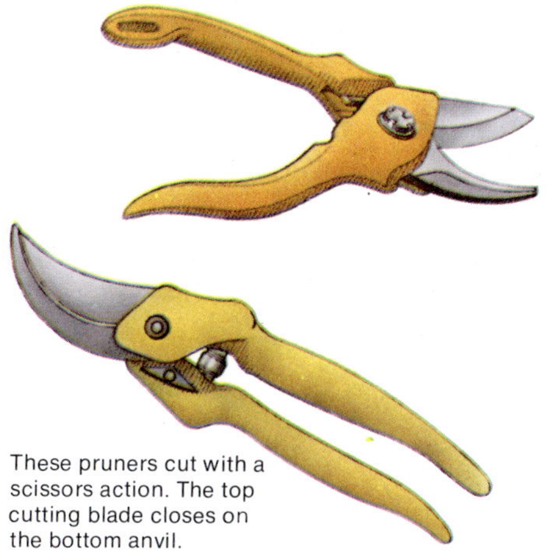

These pruners cut with a scissors action. The top cutting blade closes on the bottom anvil.

For taller trees and shrubs it is a good plan to have a special set of long-reach pruners. These have long tubular or flat metal shafts extending their handle, with a rugged cutting blade at the top. The blade is connected to a handle in the base of the shaft by a cable or rod. Some of the more sophisticated designs have extendable handles and can reach up high into a tree.

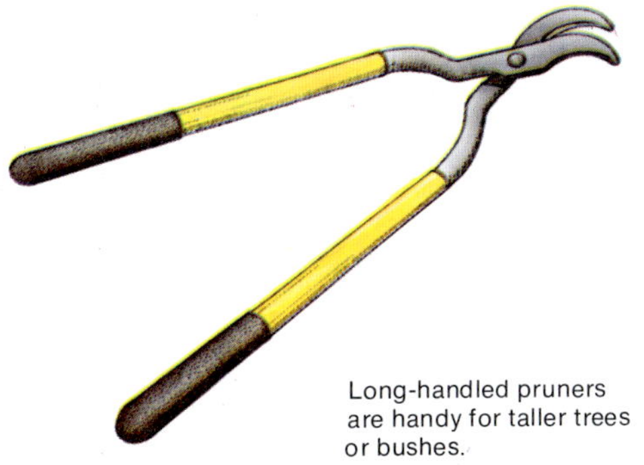

Long-handled pruners are handy for taller trees or bushes.

For tough wood, lopping pruners should be used.

There may come a time in a tree's or shrub's life when some large, tough branches must be removed. This is frequently necessary in neglected gardens. For this type of pruning special cutters, or 'loppers', will have to be used. These have short, thick handles and very strong blades. Some have a special action that gives the user extra cutting power.

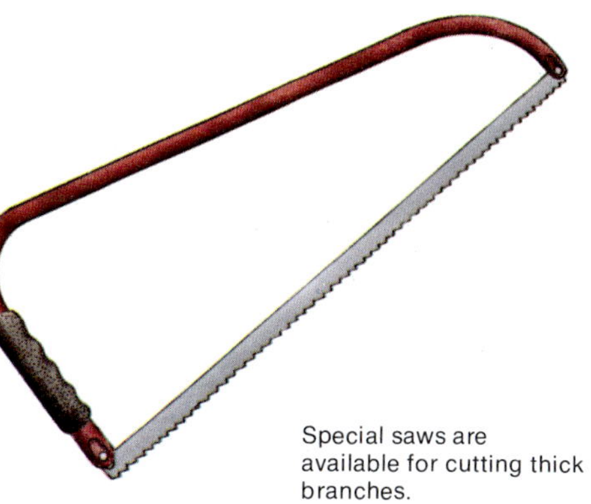

Special saws are available for cutting thick branches.

For the removal of very large limbs or branches a saw will be required. The best ones have specially-shaped handles – many of them of tubular construction with a bow shape for easy handling.

Successful pruning depends on an appreciation of what a good pruning cut is and the ability to identify various types of buds, especially on fruit trees and bushes.

Pruning cuts These must always be made close to a selected bud with the slope or angle of the cut pointing away from it. The cut should never be made so that it slopes into the bud as this can cause moisture to lodge there and rot to set in.

Buds The difference between a fruit bud and a growth bud must also be recognized. The former is quite plump whereas the latter tends to be thinner and often lies a lot closer to the stem it grows on. The fruit bud is responsible for producing the blossom that in turn becomes the fruit. The growth bud, on the other hand, makes new shoots or growths and thus extends the size of the tree or bush.

Other pruning terms Leaders and laterals are two terms with which you should familiarize yourself before pruning. A leader, as its name implies, is a leading

Laterals are growths from a main stem or leader. Fruits bear on either the tips or the spurs.

Bud identification: Plump buds (left) are the fruit buds. Thinner, pointed ones (right) are growth buds. The growth buds often lie closer to the stem.

shoot or branch of a tree or bush. A lateral is the growth or shoot that forms from the side of a main branch.

There are also the terms spur and tip. Some varieties of plant bear their fruits on short shoots or growths called spurs. Quite often, these spurs appear in clusters of more than one. Examples of spur bearers are 'Starkspur Lodi', 'Skyspur', 'Miller', 'Sturdy Spur Delicious', and 'Starkspur Winesap'. These varieties are all available as dwarf trees.

Although some plants bear their fruit on spurs, there are many that have their fruits at the tips of shoots that have been made the previous summer. These are the tip-bearing plants.

Ornamental trees and shrubs sometimes raise doubts too, but these are dealt with separately in the following pages.

Apples and pears

As with children, training apple and pear trees when they are young is of vital importance. The type of pruning which is required will depend on the age of the tree you buy. Many trees are purchased when they are one, two or three years of age. In the case of a one-year-old tree the plant is cut back after planting to leave a main stem of about 2 ft. high. The cut must be made to just above a suitable bud. From this bud, later on, and from other buds lower down the main stem, several new shoots will form. These will start to form the outline or framework of the tree.

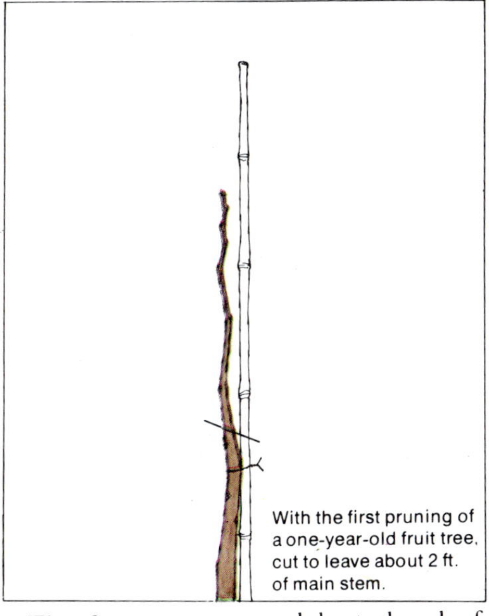

With the first pruning of a one-year-old fruit tree, cut to leave about 2 ft. of main stem.

The four strongest and best-placed of these new shoots are allowed to remain. Remove all other unwanted buds. In the tree's second year, these growths must be pruned in the winter by cutting them back to about half their length, if they are strong. If, however, they are weak, they must be cut back by about two-thirds their length. Always cut to a well-placed outward-facing bud. This will mean that the new growths will face outward, away from the center of the tree.

In the tree's third season, you will find

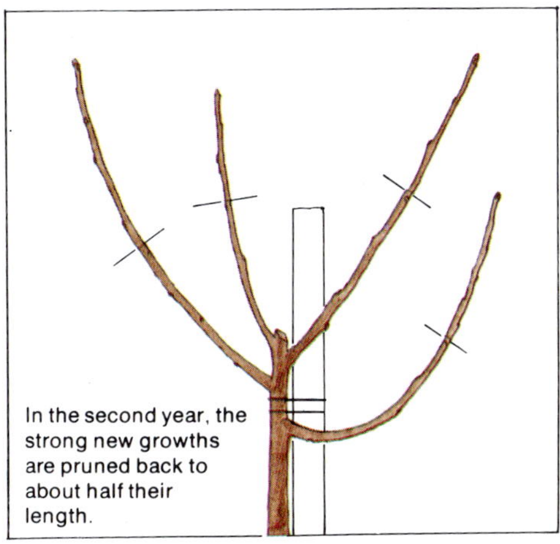

In the second year, the strong new growths are pruned back to about half their length.

that many more shoots have grown from the original ones. These will form the outline of the tree and will also become its main branches. Shorten all new shoots on each of these branches by about half their length. It will be more than likely that the tree has produced too many of these laterals so the unwanted ones can be cut back to about three buds from their base. From now on, you should have the shape of your tree.

Future pruning will follow basic rules depending on whether a tree is spur-forming or tip-bearing. Let us now take a close look at these two methods.

In the third year, more branches will form. Prune as shown for a sturdy framework.

Spur-forming trees What you are aiming for in this type of tree is to keep as much of each new season's growth as possible on which fruit can be produced. This is logical when you consider the progress of the buds. In the first year, a growth bud produces a shoot which, in the second season, produces fruit buds. In the third year these fruit buds produce spurs which in turn develop into fruit.

You must realize, however, that a second-year shoot can also form new shoots as well as fruit buds. You will therefore have a two-year-old growth which has made some one-year-old shoots. When pruning, you will want to achieve a good balance between fruit and shoot buds so that regular new growth is formed with no reduction in fruiting capabilities.

Some growths may need removal to keep the tree 'open' for easy picking. Whatever you do, you must not cut out too much one-year-old wood. Where growth is weak, you can always prune back harder to about two fruit buds.

Tip-bearing trees In this type of plant you

On a tip-bearing plant, most fruit buds occur at the tips of the new season's wood.

will find that a lot of fruit buds will be produced on the tips of one-year-old growths. Pruning is confined to cutting back to the nearest bud all those growths which can be seen not to have a fruit bud at their tips. In some cases there may not be fruit buds higher up these shoots and then a cut must be made to four growth buds from the base of each shoot.

To maintain the production of tip-bearing shoots, leading branches can be cut back to a growth bud, sacrificing the fruit buds at their tips.

Neglected trees If you have purchased a home where some of the fruit trees have grown rampant, be ruthless and do some savage pruning!

The aim must be to remove dead or worn-out branches so that new shoots are encouraged to grow and to have room in which to spread themselves. Cut large limbs out of the main trunk area. Pruning shears and a large saw are necessary for this work.

On a spur-forming tree, fruit is produced on the previous season's new growth.

First, make an upward cut into the limb to a depth of 2-3 inches. Make the cut about 6 in. away from the main trunk. Then make another cut through the branch, about 6 in. away from the previous one. The purpose is to avoid having the branch tear away the main trunk as it falls. What should happen is that the branch will crack off leaving the small stub which was previously undercut. You can then cut this off cleanly as close to the main stem as possible. Afterwards, the open wound or cut area should be treated with a sealer.

In a neglected tree there will be lots of thin, weak growth, which must be removed. Cut to a main branch each time. Slightly more vigorous growth can be cut back by about half its length. If you find badly crossed branches, remove one of them so as to avoid bad chafing.

As you prune the healthier branches, bear in mind the previous advice on tip- and spur-bearing pruning techniques.

If you want to grow an apple or pear tree in a particular shape (an attractive fan, espalier or cordon) the purpose of your pruning will be to encourage growth buds in the directions essential to maintaining the desired shape.

Pruning cuts must be made early in the year so that buds which are in the best position are allowed to grow new shoots. Sum-

Fan trees are pruned to encourage a growth of branches that can be tied against a wall or fence. Unwanted shoots are taken out.

mer pruning is necessary as well because you must restrict new growth, allowing comparatively few shoots to maintain the shape of the tree.

Generally, it is best to grow a dwarf or standard type of fruit tree in order to minimize pruning requirements. On the other hand, the advantage of fan, espalier or cordon shapes is that they can be grown in confined areas of the garden.

To remove a large dead branch, the first cut is made underneath about 6 in. from the trunk.

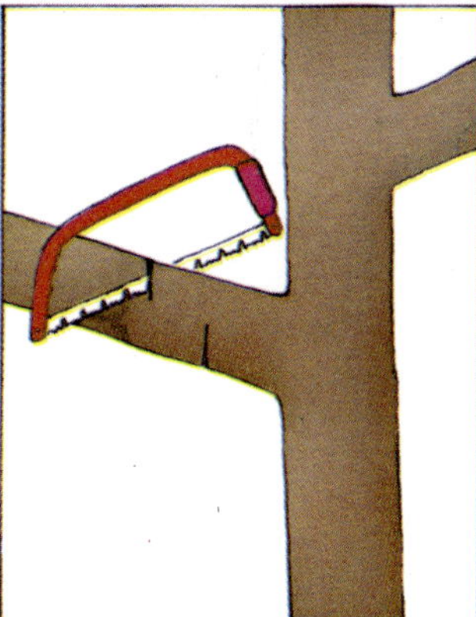

Next the branch is sawed through from above and about 6 in. further out than the first cut.

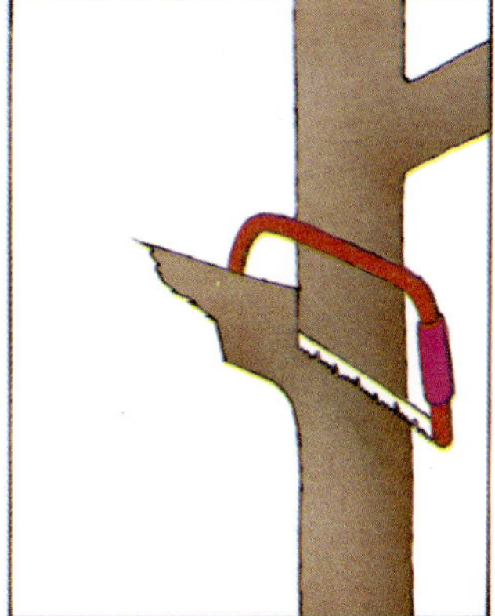

Finally, the stub is neatly sawed off close to the trunk. Treat with a sealer to prevent disease.

Small fruits

Black Currants After planting, the new bushes should be pruned back. Prune strong stems to within 6 in. of the ground. Remove any weak growth.

Subsequent pruning must take into consideration the fact that black currants produce their fruit on the previous season's new wood. Prune out some of the older growths to allow the young wood to come through. After picking fruit, remove the growth on which the fruit was borne. Always encourage young wood, of which there should be some rising from around the root area. If there is any weak young growth, remove.

With neglected bushes, remove as much old growth as possible. Try to open up the center of the bush by removing crowded growths. Left untended, these could 'strangle' each other. Cut out weak, dead or diseased shoots.

As black currants (above) produce their fruits on the previous year's wood, any older wood should be cut out to encourage large bunches of fruit to form.

Red currants (left) fruit on old and second-year growths, so as much healthy older wood as possible should be retained. Keep the centers of the bush open.

Red and white currants After planting, cut back all strong branches by about two-thirds. The weaker ones should be cut down to three or four buds from the base.

After the first year, restrict pruning chiefly to July. Remember that these currants produce their fruits on old growths as well as second-year ones. In July, shorten all the vigorous side shoots to five or six leaves. In winter, shorten the leading shoots by 2-3 in. If you see fruit buds well formed on side shoots, cut back to these.

Cut out all the dead and diseased growths in neglected bushes down to the base. Cut back weaker shoots by half their length. Try to keep as much healthy old wood as can be saved. Keep the center of the bush as open as possible.

Gooseberries The gooseberry bears its fruit on new shoots and also on spurs of older growths. Because the gooseberry has an unfortunate habit of letting its branches droop at the ends, it is a good plan to prune these back to an upward-pointing bud. Cut back vigorous branches to about a quarter of their growth and always cut to a bud position. Remove all weak shoots.

In established plants, the leading shoots should be pruned back by half. Cut out all weak shoots. Side shoots must be trimmed back to six leaves every July.

Large gooseberry fruits (left) and cherries (above) are the result of careful pruning. Trim the side shoots of gooseberries every July; cherries are best pruned during the springtime.

Cherries These are treated like apples and are usually grown on a shrubby tree. The only difference from apples is that cherries are best pruned in the spring, not in the winter. Prune after the buds have burst. The purpose of cherry pruning is to encourage the production of new growth as the cherry produces the bulk of its fruit from the previous year's growth. It may be necessary on occasion to restrict the spread of branches in a small garden.

Raspberries After planting new canes they should be severely pruned back to within 9 in. of ground level. After the first year, canes are pruned immediately after the fruit has been picked. Canes which have just carried their fruits must be cut to ground level.

If a lot of new canes have been produced and there are too many for your requirements, reduce them to about 9 in. apart. Those raspberries, which fruit in the autumn must be pruned in February. Cut canes to within 9 in. of ground level.

Prune raspberries (below) after fruiting.

Peaches and nectarines.

Peach or nectarine trees can be widely planted over the continental U.S. Dwarf and standard varieties are available, providing beauty and reliable crops of fruit to the gardener.

Training is like that described for apples and pears for those trees that are being used in a mini-orchard.

However, peaches and nectarines also do well trained as espaliers against a warm wall or fence. This allows gardeners with little space or a blank wall that needs softening with some kind of plant material, to

An espaliered peach will grow well in the shelter of a warm wall.

produce a bountiful crop.

If you have purchased a year old tree, prune as follows: cut back the main stem to about 2 ft. from ground level to just above a bud.

Shoots will be produced from the buds. When they are about 1 ft. long, tie them to training wires, angled at approximately 45 degrees. Do not allow any buds or shoots to develop below these selected shoots.

Allow two strong shoots to mature and in the second year cut these back to just above

Bamboo canes and wires are used as supports to train branches of a young tree at 45 degree angles.

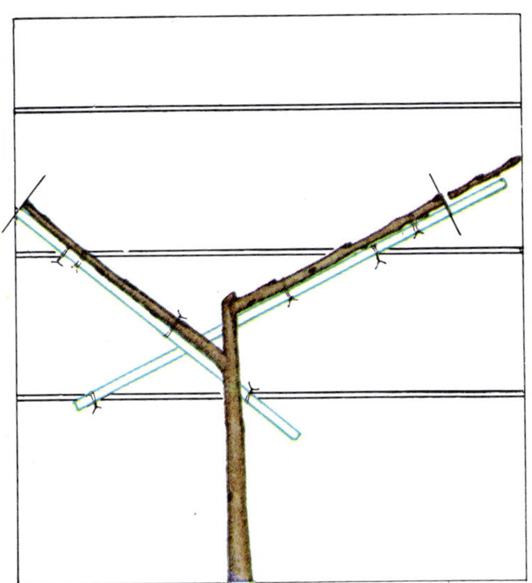

a growth bud and to a length of 18 in. From these shoots, extension shoots will be produced. These should also be tied to the wires.

New shoots will develop on the extensions. Two should be allowed to grow on the top side and one or two on the lower side. By tying these inward, you will add to the fan shape. The next winter, prune these growths back to a growth bud and a length of about 2 ft.

Later in the season extend the tree by training the leading shoot from the end growth bud on each shoot along the wire. At the same time, train in two or three top-side shoots, and two from below it, to fill in wall area and build up a more dense fan arrangement.

Try to achieve a well-balanced fan by making sure that all shoots are evenly spaced along the main framework of the tree as it grows.

Once your espaliered tree is produced, subsequent pruning in autumn requires cutting out old laterals and training new ones. This is best done after fruit has been gathered. In May and June light pruning of unwanted shoots can be done, but take only a few shoots at a time.

The purpose of pruning an espaliered tree is to produce evenly spaced shoots over the widest possible surface.

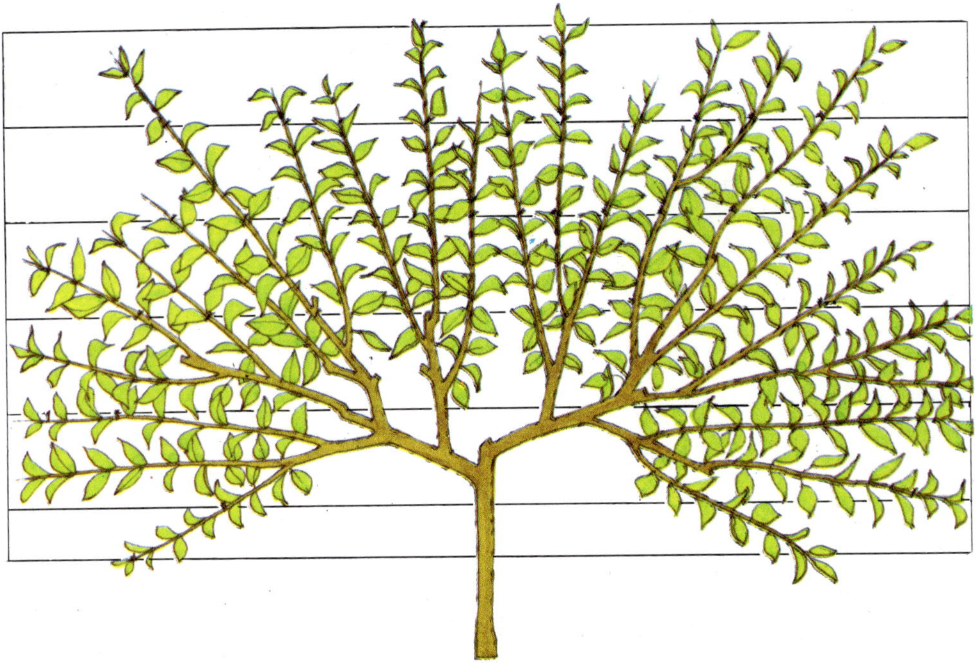

Blackberries and loganberries These are invaluable cane fruits which take up little room in a garden because they can be grown against walls or fences. Pruning is very simple. After new canes have been planted, they should be cut down to about 12 in above ground level, making the cut to just above a bud.

Later on after picking, those canes which have just carried the fruit should be cut back to ground level. The new canes are then trained neatly in their place.

Grapes A new grape vine can be planted in the summer or the autumn. Just before the foliage begins to fall, the first pruning is undertaken. The summer-planted vine will have made more growth

Blackberry canes which have fruited are cut out. Only the new canes produce more fruit.

Strong. vigorous growths are the ones you should encourage for the production of heavy bunches of grapes.

than an autumn-planted one so it should be cut back once it has grown to about 9 ft. For the autumn-planted one the cut is made when the vine has attained a length of about 4-5 ft. These length suggestions can only be approximate as growth varies a lot according to variety, weather conditions and, of course, the care vines have received.

Always make a cut to just above a plump bud. The amount of growth you cut off will depend on how many new growths you wish your vine to produce. The harder you prune, the more growths you should get. Cutting back by two-thirds of the length will give a good, vigorous single growth. A harder pruning to one-third the length will make the vine produce several shoots.

In the spring of the following year, the one-stemmed vine should be allowed to grow on its leading shoot. All others must be cut off carefully. For a multi-stemmed vine, allow about two nice strong leading shoots to grow and remove all others.

Do not allow the young vine to produce flowers the first season—remove them.

In subsequent seasons, your vine can be considered an established one, and the autumn pruning after leaves have all fallen will consist of cutting back new growth from the leading shoots by about one-half their length.

Any lateral should be cut back to two buds. This pruning routine will build up a really strong vine.

Plums and damsons For bush and standard types of trees, the pruning is as for apples, with the aim of building up the frame of the tree. Afterwards, the leaders should be cut back or shortened a little each March. Where the tree is a variety that has a rather drooping type of growth, the cut should always be made to an upward pointing bud. Where the tree has an upright habit, make the cut to a bud which is pointing outwards.

Where there are growths which are weak, cut them right back to their

Pruning techniques for plums and damsons same as for apples.

For trees with drooping habits, prune back to an upward-pointing bud. If upright in growth, prune to a downward-pointing bud.

Wisteria is a beautiful rambling type of shrub. Prune in July and also in November.

base in late summer once the fruits have been picked.

If you have decided to grow an attractive espaliered tree your pruning sequences are as follows. Prune a one-year-old tree (often referred to as a maiden) back hard to about 18 in. from the ground in November. To form the fan shape retain four strong shoots

which are produced the following spring and train them against a guide wire (or other means of support), spaced out to form a small fan.

In November of that year, cut these growths back to about 18 in. lengths each. Each spring thereafter, continue the thinning and retraining process. During each growing season the tree will fill in the required space, and by neatly training the many shoots or branches a very attractive fan tree will be produced. Once the shape and size has been achieved, prune regularly as for an established tree.

Shrubs

Usually, shrubs need little pruning, except to maintain their pleasing shape or to remove dead or diseased wood. Where a neglected garden has been taken over it may well be necessary to cut back shrubs ruthlessly if a lot of weak, straggly growth has been produced. The centers of the shrubs will most likely be very over-crowded and some drastic thinning in the area will need to be carried out. Crossing and badly-placed branches will also need to be removed.

It is useful to know that some shrubs produce their flowers on new growth and others on the previous year's wood. In the case of the former, spring treatment is best, cutting all the previous year's growth back to three buds from the base.

For the latter type of shrub, pruning is carried out only after flowering. Seek out the branches which have just borne their flowers and cut back to about three buds from the base. There are some vines, such as clematis, which need 'sorting out' as far as their rather complex growth is concerned. This plant can get out of hand over the years unless much of the growth is cut back. For the large-flowered hybrids late winter treatment is required, and for those varieties which flower twice in a season, cut back

Clematis easily gets out of hand. Its rather complex growth must be regularly pruned to keep it in order.

March and April are the months for pruning climbing roses. Strong shoots need cutting back by a quarter.

Floribunda roses (above) and hybrid tea roses (left) are both pruned while dormant, but after the danger of frost is past.

lightly after the first blooming period has finished.

Roses

New roses. Plants that generally require no pruning after planting are ramblers, climbers and species types, unless there is damage to canes. In the second year, remove about 10 percent of old wood.

Because packaged, containerized and bareroot hybrid teas have already been pruned to 3-4 heavy canes, further pruning should be done only to remove dieback, a brown area that may appear on the top of a cane. Prune back to an outward growing leaf bud in the green part of the cane. In the case of bareroot roses, also check roots, pruning those that are broken.

This same treatment should be followed with floribundas. Prune polyantha types back to about a third of their original height.

Established roses Shorten the young shoots of hybrid teas to four buds. Medium-vigor shoots should be cut back to two buds. Do the work in spring, late spring being better in cold areas.

With floribundas study the different types of growth on each plant, distinguishing between strong , medium and weak-vigor growths. For strong growths, cut back to about six buds; for medium growths cut to four buds, and for the weaker ones prune to one or two buds. The latter treatment should encourage the production of sturdier shoots later in the year. Prune in spring.

Prune polyantha roses in spring also. Cut out all weak or thin shoots. Cut back other stems to about half their length.

Climbing roses must have all their old or diseased growths cut out in spring. Then cut the young branches back by approximately two-thirds their length. Prune the climbing 'sports', as they are called, in spring or late spring, cutting strong shoots back by a quarter, medium growths by a half, and weak ones by at least two-thirds in order to encourage stronger growths.

Prune shrub roses from autumn through to spring in mild climates. Elsewhere wait until spring. Thin out badly-placed growth and dead or diseased shoots at this time too.

Tree roses usually require very little pruning. For the most part pruning can be confined to the removal of dead or diseased branches and of any growths that are so badly placed that they spoil the shape or appearance of the tree. An open center to a tree – especially the round-headed types – should be the aim. For this reason some thinning out of the central branches may be required from time to time.

Trees

Deciduous trees (those that lose their foliage in the winter) should be pruned after their leaves have fallen. Evergreens (those that do not lose their leaves) should be pruned, if necessary, in spring.

Climbing roses (above) should have old and diseased wood cut out and thin wood removed. Make all cuts back to a good bud. Remove any old flower stems.

Cutting lines show how to prune an established hybrid tea rose (left). The extent of cutting depends on the vigor of each shoot. Remove dead and diseased wood.

6 Weeds and weed control

It is a great pity that weeds grow just as well as other plants in the garden. In fact, many weeds grow a lot *better*, and thrive in even the poorest growing conditions. Because weeds grow so prolifically, careful organization is needed so that they can be controlled with the minimum of effort and time.

Unless weeds are promptly dealt with, they will quickly seed themselves, thus adding to the misery of keeping them out of the garden. Weeds are generally vigorous

Care must be taken when applying weedkillers. They can be sprinkled onto vegetable plots by watering can or spread on lawns by special spreaders.

Weedkillers

Fortunately, there exists a wide range of chemicals that, if used sensibly, will greatly reduce the effort required in the battle against weeds.

A number of sensible and important precautions must be taken when using weedkillers. Keep them well away from children and animals. Always follow the instructions to the letter and always wash out containers thoroughly after use. Never use the same watering-can for ordinary watering as well as weedkillers. An *old* can, and one marked as such, should be kept especially for this purpose.

The use of weedkillers can be so complicated and fraught with risk for surrounding vegetation that many gardeners are returning to traditional methods of control: mulching and cultivation.

Chemicals do remain useful for eliminating weeds from lawns and paths and driveways, although in the latter case care must be taken that chemical run-off does not reach nearby growth and kill it too.

Lawn weedkillers are usually in dry form and applied with fertilizers by a spreader. Wax bars and spot weedkillers, such as the Killer Kane, can be obtained.

plants and will quickly smother crops if allowed to grow. Of course, they vie with other plants for food and moisture, often to the detriment of these plants, and ruin the garden's general appearance.

Lawn weeds

Chickweed This annual weed reproduces by seed. It has tiny heart-shaped leaves under 1 in (3 cm) in length. Use mecoprop or, for a recently made lawn, a weedkiller specifically recommended for new lawns.

chickweed

Cinquefoil This is a perennial – in other words, a plant that continues to live for many years and that, in many cases, flowers time after time. Cinquefoil has a five-leaf pattern of a serrated form. Flowers are yellow with tiny reddish stamens. Cinquefoil seeds itself. Several applications of mecoprop should keep it in check.

clover

cinquefoil

Clover Many lawns look lovely and green from a distance even in drought – but a closer inspection often reveals the fact that most of the growth is clover. This plant is a perennial with a three-leaf pattern formation (occasionally four-leaved). Clover has creeping stems and so spreads very rapidly. It has white ball-like flowers and also seeds itself. Use mecoprop or dicamba to control clover, and feed the lawn with a high nitrogen fertilizer in the spring.

Creeping buttercup A perennial that, as its name implies, spreads quite rapidly by its creeping stems. It also seeds itself. It has triangular, serrated leaves and bright yellow flowers. Treat it by watering on a 2,4–D mixture. Dicamba is also used.

creeping buttercup

Buckhorn Perennial found in many lawns that have been uncared for. It has tapering leaves and rather insignificant brownish heads. Buckhorn seeds itself. Control it by using mecoprop or 2,4–D.

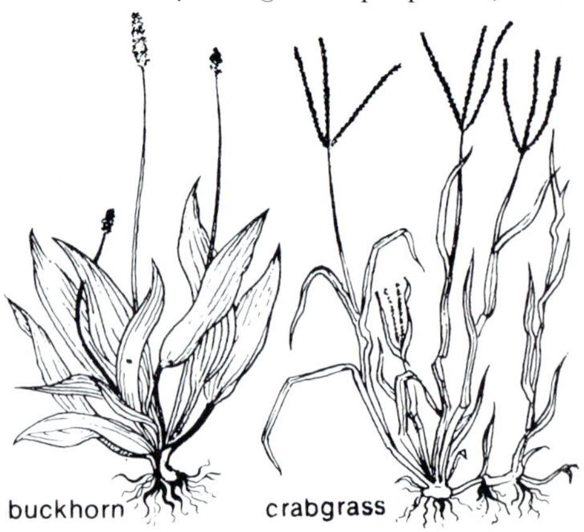

buckhorn   crabgrass

Crabgrass An annual grass weed common over most of the country, and a special nuisance in lawns. It is a creeping grass with rather wide blades. The best control is a pre-emergent chemical to prevent the seeds from germinating. Siduron (available under trade name Tupersan) has proved effective if applied at the right time in spring and used according to directions.

Dandelion This perennial has serrated leaves and bright yellow flower heads. The plant reproduces itself freely by seeds. Even if a new plant is cut down, its fleshy tap root will soon produce a new growth. Repeated application of mecoprop or 2,4–D should be effective in keeping down dandelions.

dandelion

Pearlwort This perennial is a very common invader of lawns. It seeds a great deal, has very thin narrow leaves on spreading stems and tiny yellow flower heads. Use mecoprop, dicamba or silvex to control it.

pearlwort

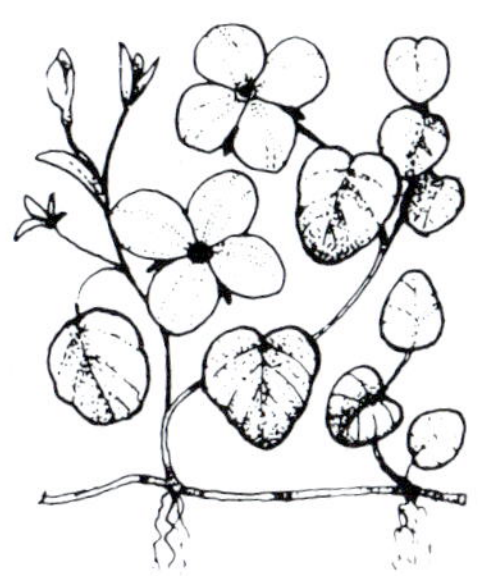

speedwell

Speedwell This is a perennial weed often found in lawns. It has pretty, tiny blue flowers and small, heart-shaped foliage. It takes root along its long stems and is generally quite fine in growth. Apply DCPA (a trade name is Dacthal) and use according to instructions.

Wild garlic or onion Also known as field garlic, this bulbous perennial can disfigure lawns with its tufts of typical onion-like leaves. Seeds itself and is also easily spread by its underground bulbs. Use 2,4–D with a wetting agent as a spray or use a wax bar such as the Ortho Weed Bar. A very persistent weed.

yarrow

wild garlic

Yarrow Another perennial, this plant has ferny-type leaves and tall heads of tiny white flowers. Several applications of mecoprop will control it. Sulfate of ammonia feedings will also help reduce infestations of this weed. Dicamba can also be used.

Flower and vegetable plot weeds

While you have been keeping an eye on your lawn weeds, other weeds have probably been settling in elsewhere in the garden. Let us take a look at the more common types that you may discover in your flower and vegetable plots.

Bindweed   A really nasty weed that can really take over a garden, bindweed climbs to fantastic lengths and smothers plants with its heart-shaped leaves. It can be recognized by its trumpet-shaped white flowers. Bindweed is a perennial that comes up early each year. It is easily spread by its fast-rooting underground stems. Concentrate the weedkiller application around groups of plants. Spot treatment, on the other hand, is a somewhat tedious but very effective method of attack. To spot-treat, take a paintbrush and dab concentrated weedkiller onto it. Use 2,4–D for this.

Couch or quack grass   This is another nasty perennial weed that is difficult to deal with if it is allowed to get out of control. It makes life very difficult for a gardener, especially if he is taking over a new or neglected garden. It has broad, pointed leaves and green heads. Use dalapon to control it. Quack grass also has masses of creeping stems that can easily be rooted out before planting. Deep digging and then forking out and burning of roots is another method of dealing with this weed.

Dock   The broad-leaved dock, which is a perennial, has long spikes of brown heads. It seeds only too easily. If the long tap root is cut off but left in the ground, this too will grow again and continue the infestation. Take your time and patiently spot-treat the weed. For large infestations, however, it is better to apply a watering or spray of dichlobenil, dicamba or silvex. Spot treatment should be done with 2,4–D.

Groundsel   An annual weed that spreads by its seeds, groundsel has serrated foliage and bunches of yellow flower heads. There are many types of groundsel, which are all members of the Compositae family, also known as the sunflower family. Groundsel grows on dry plains and hillsides, particularly throughout the mountains of California and Oregon, where it blooms in the autumn. It is spread by the wind, and can be dealt with by applications of dichlobenil.

Nettle   This plant is a real nuisance. Not only is it a perennial that seeds easily, but it has stinging oval leaves. Several applications of 2,4–D or 2,4,5–T are effective control measures. A spring application is always a good idea.

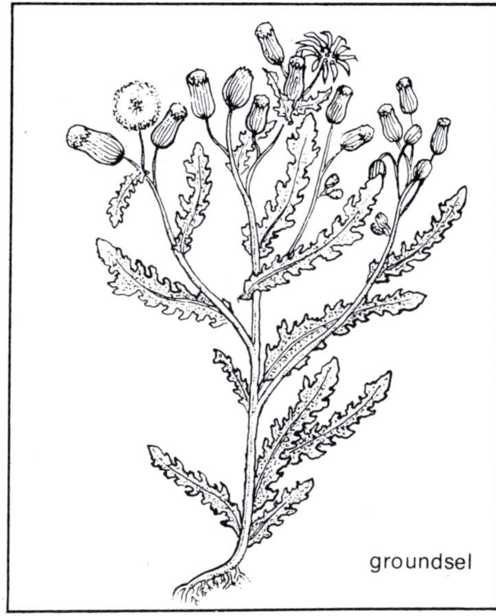

Oxalis   A perennial that soon spreads by means of the tiny bulbs at its base, oxalis also seeds a lot. It has clover-like leaves and tiny pinkish flowers. Try 2,4–D plus silvex, following directions on container. Be careful when digging over the soil, for that's when the little bulbs break off and scatter. Try to remove it with surrounding soil.

Horsetail   Another quite nasty perennial weed with fir-tree-like growths that start off as closed spikes of brownish shoots, horsetail is a non-flowering weed which gets its name because of its resemblance to a horse's tail. It spreads quickly by its root system. Usually growing on dry land, horsetail is able to spread downhill rapidly and is very difficult to eradicate. Repeated applications with dichlobenil will check it, although it is a very difficult weed to deal with and unfortunately there is no foolproof answer.

Poison-ivy This woody vine, with its compound leaves composed of three leaflets, is widespread over the eastern half of the country. A close relative, Pacific poison-oak, grows on the West Coast. Both are easily spread to home gardens by birds. All parts of the plants can cause a painful skin rash. Eradication is difficult. Both ammonium sulfamate (Ammate) and 2,4–D plus silvex are recommended.

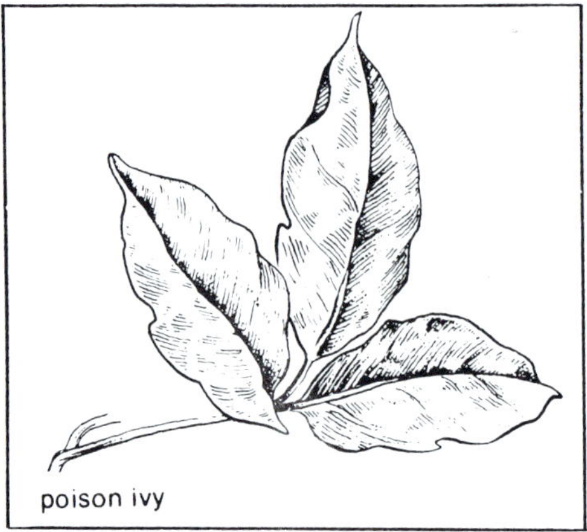

Plantain A perennial plant with broad leaves and spikes of brown heads, plantain seeds easily. It is controlled by mecoprop or by using 2,4–D.

Shepherd's purse This plant is an annual with deep-toothed leaves and tall greenish heads. It seeds all too easily. Apply successive treatments of dichlobenil (trade name Casaron) to control it.

Thistle There are various types of perennial thistle, and they can be a nuisance in the garden, for they not only spread by seed, but also by vigorous creeping underground stems. Thistles have long serated, prickly leaves and reddish or purple flowers. They should never be allowed to flower and roots should be removed when digging. Control with dichlobenil.

7 Controlling pests and diseases

One of the 'wars' all gardeners have to wage from time to time is against the pests and diseases that can attack your plants. The type of attack and its severity will vary considerably from season to season.

There are several pests and diseases that can be termed 'common' to the garden. It is well to be able to identify them early so that the best methods of control can be put into practice. Most pests and diseases you'll come across will be found in this book under the crops they may attack.

Spraying There is a lot to be said for taking precautionary measures in good time – even before an attack has begun. A regular spraying program can prevent

A powerful mist or spray is needed for trees so that a good penetration is assured.

A hand sprayer is useful when spraying smaller plants such as rose bushes.

troubles. With some types of preparation that are assimilated by plants early spraying can fortify the plants and make them reasonably resistant to attacks.

Successful control of pests and diseases also depends on spraying or dusting when conditions are favorable. Choose a windless day so that the material being applied is not blown all over the place. Care must be taken to reduce the drift of spray or dust to a minimum to prevent contamination of food crops (and your neighbors' gardens).

Control of pests and disease will also depend on the use of a good sprayer when liquid preparations are used. A forceful spray is essential in order for it to penetrate well into foliage, branches and flowers.

In a large garden, a pressure sprayer is ideal, for this can be pre-pressurized with a few priming pumps of the handle so that a continuous powerful spray is emitted. This saves a lot of time and effort. For the smaller garden, there are several excellent small-capacity sprayers, some of which can be pre-pressurized.

Successful control of pests and diseases also depends on spraying or dusting when conditions are favorable. Choose a windless day so that the material being applied is not blown all over the place. Care must be taken to reduce the drift of spray or dust to a minimum in order to prevent contamination of food crops (and your neighbors' gardens).

Safety *Read the instructions carefully*, and follow them. Keep all preparations well away from children and pets. The best place is in a cupboard which can be locked up, or on a high shelf.

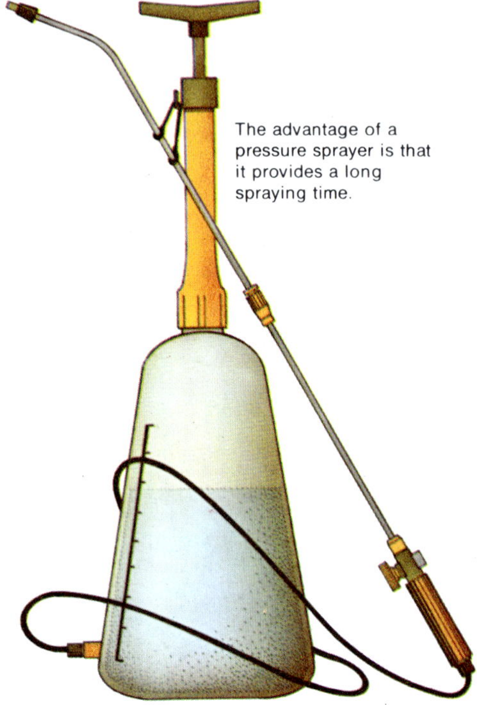

The advantage of a pressure sprayer is that it provides a long spraying time.

Diseases

A wide variety of diseases may also affect garden plants as well as those grown inside greenhouses and frames. Controlling diseases requires a sound knowledge of their symptoms.

Keep all chemicals safely locked away from children and all tools carefully stored away from their reach.

Black spot As its name implies, this disease is indicated by a distinct black area or spot on the foliage. Also, leaves turn yellow and fall off. Roses in particular are prone to black spot. The worst attacks occur from June onwards. Yet, the disease can be controlled if foliage is not wet at night.

Pick off and burn affected foliage as soon as the disease is noticed. Spray with captan, maneb or zineb.

Blight Potatoes are subject to this disease. Blight is first revealed by the foliage acquiring black spots and patches. The stems are also affected and the tubers themselves will have areas of brown decay on them. Outdoor tomatoes may be affected as may those grown under glass if they are near an infected potato bed.

Blight is usually first seen in late June or July. However, spraying with Bordeaux mixture before this time should do much to prevent attacks. Maneb or zineb are two other excellent control products.

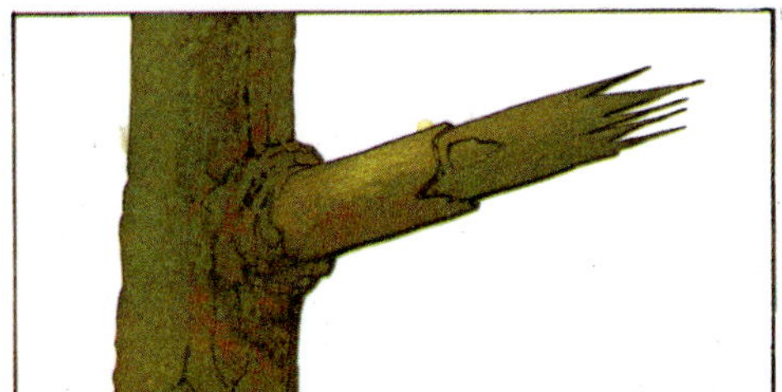

Canker Swollen cankered areas on the shoots of apple-tree trunks are a sure sign of canker. This disease can be troublesome at any time. It is necessary to burn pruned-out branches and larger growths wherever possible. To prevent canker, spray with Bordeaux mixture before leaf-fall and also at bud-burst time.

Parsnips can be affected by a canker as well. The shoulder or top part of the root becomes covered with a reddish-brown rot. The trouble can be spotted at any time during the various stages of a plant's growth.

The only sure means of control is to pull up and burn badly infected specimens. Rotating the crops and growing more resistant varieties is also a good means of controlling this.

Black spot (left) affects roses particularly severely.

Blight affects the foliage, stems and roots of potatoes.

Decaying wood and peeling bark are sure signs of canker on an apple tree.

Chocolate spot This is a disease which affects broad beans, especially those which were sown in the autumn and overwintered. The signs are usually noticed in late December on winter plants and in June on spring-sown ones.

Small dark brown areas or spots are seen on leaves and stems. Fortunately, it is not necessary to do much spraying as long as the plants can be kept growing as strongly as possible by applications of sulfate of potash at not more than 1/2 oz. per square yard.

The secret of avoiding this disease is to make the young plants as strong and healthy as possible by ensuring sufficient potash is in the ground at the time of sowing.

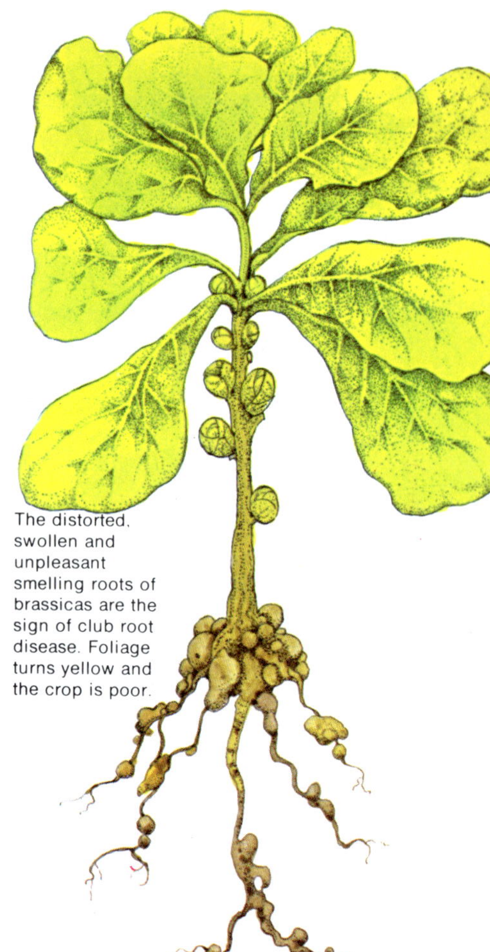

The distorted, swollen and unpleasant smelling roots of brassicas are the sign of club root disease. Foliage turns yellow and the crop is poor.

Small brown areas on broad beans are a sign of chocolate spot.

Club root Most of the brassicas—the cabbages, Brussels sprouts, turnips, wallflowers and cauliflowers—can fall victim to this disease. The symptoms are swollen and distorted roots and weak, yellowing foliage.

The only really effective solution is to pull up and burn affected plants. Liming the soil does, however, reduce the likelihood of trouble. At the beginning of the season place some club root dust in each of the planting holes and also dust the seed bed well. Roots may also be dipped in a calomel dust paste or 'mud' and then planted out. Rotate crops so that brassicas are never planted on the same ground without a three to four-year interval.

Damping off Seedlings are especially susceptible to this trouble. The young plants rot at ground level and then fall over. To prevent damping off, use only clean (sterilized) soil mixtures. This is why it pays to purchase ready-bagged composts, which contain sterilized soil.

The disease can also be checked by applying Truban on floral crops and Terrachlor WP to vegetables.

Damping off (above) causes young seedlings to fall over, wither and eventually die.

Cabbages are especially susceptible to club root, but liming the soil will increase the chances of producing healthy specimens like the one below.

The above lettuce is happily free of grey mold, having been grown in a greenhouse with adequate ventilation.

A greyish rot over the leaves of greenhouse chrysanthemums (below), tomatoes and lettuce is a sign of botrytis attack.

Botrytis Known also as grey mold, this can affect many plants but is especially troublesome in greenhouses. Tomatoes, chrysanthemums and lettuce are three of the most common plants to be attacked.

A greyish mold grows over the foliage which also rots. Plants are affected all year round. It is necessary to burn infected foliage at once. Also, make sure plenty of ventilation is available in the greenhouse as this will help to reduce the trouble. The best defense is a systemic fungicide such as Benlate.

Leaf mold This is often found among greenhouse tomatoes. The attack is seen as patches of purple-brown fungus or mold on the undersides of the leaves. Yellowish areas or blotches will be noticed on the leaf surface. Spray with systemic fungicide, and try to avoid a close, humid atmosphere in greenhouses by providing more ventilation. Better still, grow only those varieties of tomatoes which have been specially bred to resist the fungus. April to May and again in early June are the most likely times of attack.

Mint is one of the plants susceptible to rust disease.

To avoid leaf mold damage, resistant varieties of tomatoes such as this should be grown.

Rust Masses of tiny brownish spots on the foliage and stems of plants is a sure sign of rust. Hollyhocks, roses and mint are especially susceptible to this disease, which affects plants during most of the year.

Rust can be controlled by picking or cutting off as much of the affected growths as possible and burning them. Spray with zineb at ten-day intervals for plants in greenhouses or frames. Use thiram or liquid copper fungicide frequently.

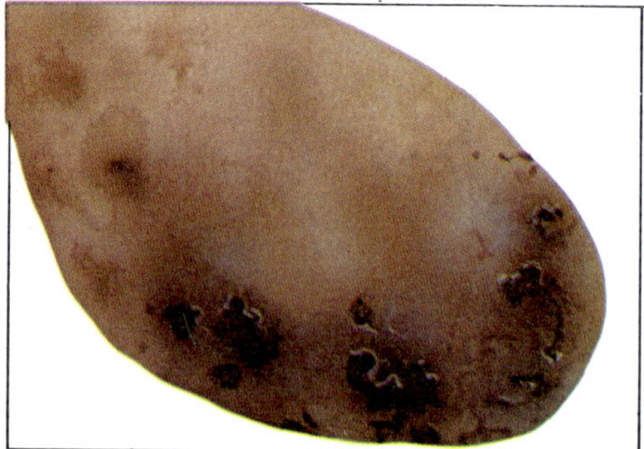

Scab This is the common type of scab disease which attacks potatoes. The skins become covered with irregular scabs. It is necessary to burn affected tubers and also to avoid planting on the affected plot for at least four years. Avoid the use of lime for potatoes as this will also encourage the formation of scabby skins. A good idea is to plant the more scab-resistant varieties of potato.

Verticillium wilt This disease is very common in tomatoes grown under glass. The whole plant wilts during the daytime and looks as though it badly needs water. The plant seems to recover during the night only to be in distress again next day.

Remove and burn badly affected plants while keeping the others growing as strongly as possible with a top dressing of moist peat around the base of the stems. This should encourage new roots to grow that may save the plants.

At the end of the season the soil must be cleaned or sterilized with a weak solution of formaldehyde, used strictly according to the instructions. No other plants must be in the greenhouse while this work is carried out. The problem can be largely eliminated, however, by growing verticillium resistant tomato varieties.

Top-dressing a tomato plant (right) with a layer of moist peat may save it from destruction by verticillium wilt disease because it encourages new roots to grow.

Knobby markings on the skin of potatoes (left) are a sign of scab disease. By comparison healthy potatoes have a taut, shiny skin (right).

Virus All types of plants can be affected. In most cases, growth becomes stunted or dwarfed. Foliage may become twisted and distorted too. There is no cure and so infected plants must be lifted and burned as soon as possible. Healthy plants will be more resistant than weak ones, so purchase only good-quality seeds and plants.

The preceding list covers most of the more common pests and diseases which you may encounter in the garden. In many cases early preventive measures can be taken, which will always pay off. However, one must be ruthless with diseased plants. Burn them at once if they become infected.

Regular spraying with fungicides will control common diseases such as those illustrated in the insets: (from left to right) gladiolus scab, bean rust, and strawberry leaf spot.

Plants infected by any virus disease should be burned at once. This should halt the spread of the virus. A sign of severe attack is the twisting and distortion of foliage.

Aphids can be controlled under glass if a special fumigating 'bomb' is used.

Plants can become quickly and badly affected by attacks of aphids. The insects weaken plants by sucking the sap.

Pests

Aphids These are perhaps the most common garden pests and they affect a great many plants. What you will see are masses of tiny insects—white, green or black. Most will not have wings, although a few may. As well as garden plants they will also attack greenhouse crops such as tomatoes and cucumbers. The white fly is especially troublesome here.

Aphids suck the sap of plants. After bad attacks the plant becomes weak and the foliage discolored and yellow. They also attack buds, and poor growths result. Attacks start in the spring and continue throughout the summer. Under glass, aphids—especially the white flies—can be present at any time of the year.

Frequent spraying, especially at the very first signs of insects, will reduce the attacks and might control them completely.

Under glass, an aerosol, such as Whitemire, can be used. Make sure the greenhouse door is shut and that all cracks are temporarily sealed so that most of the fumes are contained inside the building. Afterwards, roughly twelve hours after the aerosol or fumigant has been applied, the door and ventilators must be fully opened to release any remaining fumes. You must *not* remain in the greenhouse after the fumigation has been started and instructions must be very carefully followed. Certain preparations cannot be used for crops such as melons and cucumbers.

Apple sawfly From May to June the young fruit of apple trees may be attacked when the caterpillar of the sawfly eats its way into their cores. Later, in the mature fruit stage, you may see surface scaring.

The best control is to spray trees with a multi-purpose fruit tree insecticide immediately after petal fall.

The apple sawfly caterpillar eats into the fruit and to the core where its activities cause rot to set in.

Big bud This is a very explicit description of a trouble which can affect black currants. If dormant buds in winter are seen to be badly swollen, the best method of control is to pick them off and burn them or to cut out badly affected stems entirely. Red and white currants can similarly be affected, as can gooseberries on occasion.

Spray in the spring when the early leaves are about 1 in. in diameter with lime-sulfur using twice the normal winter strength.

The caterpillars of the cabbage root fly devour the plant's roots.

Cabbage root fly It is usually young cabbage, cauliflower and Brussels sprout plants which are affected. Plants flag or droop badly because their root systems have been eaten into by white maggots. April to September are the months when this pest can be most troublesome.

Pull up and burn badly affected plants. Thuricide is an effective control as is malathion.

Capsid bug This pest unfortunately likes a wide range of plants. Leaves look as though they have been peppered with shot after an attack of capsid. Control is achieved by spraying with malathion or a systemic insecticide as soon as damage is noticed. The period of attack is from April to early September.

Chrysanthemums, dahlias and several other flowering plants are all prone to attack by this pest which causes buds to open lop-sided and misshapen. Trouble is worst from June to about late October, while under glass the capsid may continue its attacks even later.

Close-up of the carrot fly grub which causes so much damage to valuable root crops. Roots will become badly tunnelled.

Carrot fly   Roots are tunnelled into forming brownish-black 'cracks' which soon begin to rot. The tunnelling is done by the maggots of the carrot fly. These are prevalent from about June to late October.

Pull up and burn any badly affected roots. Thuricide or rotenone are effective

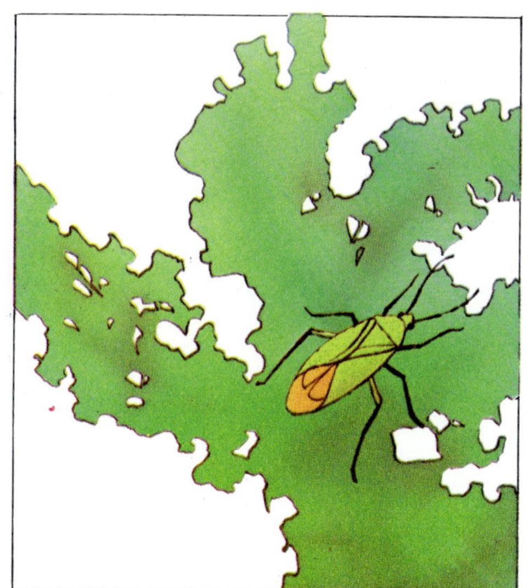

'Peppered' leaves are an indication of attack by the capsid bug (above). Dahlias are especially prone.

controls.

Sometimes, attacks of the carrot fly can be avoided if carrots are not sown until the end of May.

From late March onwards, keep a lookout for the typical signs of leaf damage caused by caterpillars.

Caterpillars   If you notice leaves with irregular-shaped holes eaten out of them, it is more than likely that the plants have been attacked by caterpillars. The attacks usually occur from late March onwards.

Use thuricide or sevin dust to control these creatures. In small attacks, and if you can spare the time, pick off and destroy caterpillars as you notice them. Look under the leaves as well as on top.

Chrysanthemum eelworm   The first signs of an attack by eelworms are outbreaks of brownish patches around the veins of the lower leaves. As the attack proceeds, leaves drop off and the flowers

shrivel. Eelworm infestations are worst in warm, moist conditions. Control by burning the infected plants and by allowing the soil to stand empty over winter.

The grub of the codling moth eats into the core of fruit.

Codling moth Pears and apples can have their fruits ruined by the attentions of the codling moth caterpillars. These eat into the fruits right to the very core. June to August are the most likely months for trouble. A malathion spraying in mid-June and a repeat three to four weeks later should do much to control this pest.

A beautiful display of chrysanthemums is well worth the effort of good cultivation and pest control.

Cutworms Shoots eaten through at soil level or stems partly eaten into are signs of an attack by cutworms. Young seedlings in the vegetable garden are particularly prone to these attacks. The cutworm resembles a fat caterpillar. It will be found in the soil around the base of the plants.

Dust soil around plants, especially at planting time. Attacks are most prevalent in the early spring and late summer months. Use sevin, thuricide or rotenone to achieve control.

The cutworm attacks plants stems at soil level.

Earwigs Half-eaten petals are a sure sign of trouble with earwigs. Many flowering plants are attacked by them, especially dahlias. Late May to October are the months when earwig attack is most likely. The best control is sevin dust.

Earwigs can cause a lot of damage to flower petals. Dahlias are particularly susceptible to attack.

Leaf miner White, twisting tunnels, particularly in the leaves of chrysanthemums, are sure signs of attack by this pest. Quite often, the tiny miner can be seen between the tissues of the leaves and can be killed by squeezing. The best control is by removing and burning all badly affected foliage and to use a spray of malathion or diazinon. Attacks occur from July to December.

White tunnels in leaves are signs of the leaf miner.

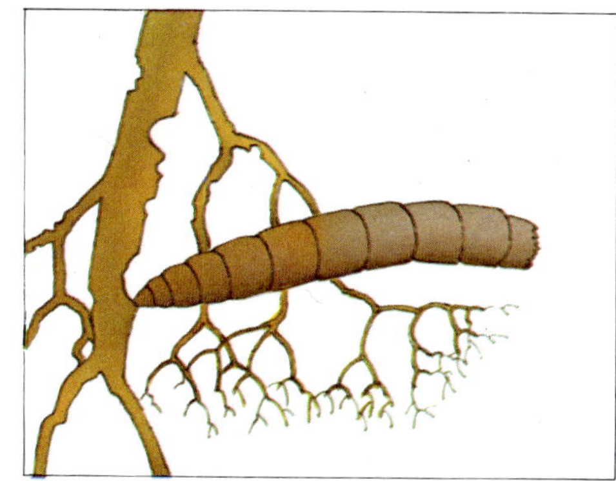

Leatherjackets feed on the root systems of plants.

Leatherjackets Many vegetable crops as well as ornamental plants can be attacked by these pests. The root systems are eaten by fat, greyish legless grubs with very tough skins—hence their name. Plants begin to droop or wilt badly and eventually die if too great a proportion of their root system is infested.

April to June are the months when plants are most likely to be attacked. These insects can be controlled by pulling up and destroying badly affected plants and using sevin dust in soil that is around the plants.

Onion fly Not only does this pest damage onions, but it will also affect leeks and shallots. The bulbs become rotten and many small white maggots infest the tissues. May to August are the danger months.

Pull up and burn badly affected crops. Take early precautions the following year by dusting the planting site well with calomel dust. On plants, spray with diazinon.

To obtain pest-free peas, such as the above, be careful to apply insecticides at exactly the right time.

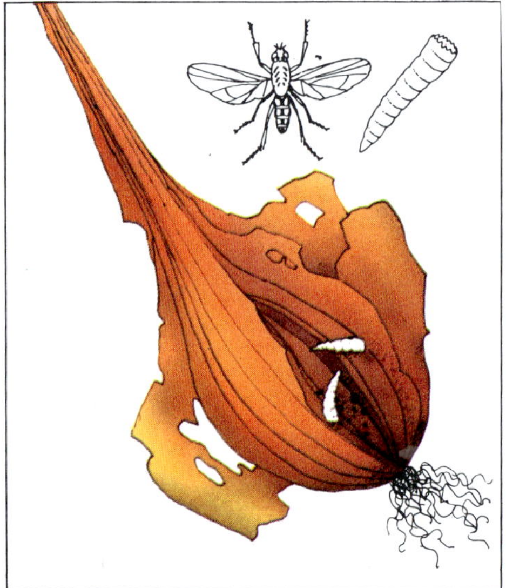
White onion fly maggots damage the flesh of onions so that rot easily sets in.

Pea moth If you have ever opened a pod of peas and discovered the peas eaten into by several small white maggots you will have confronted the pea moth. These pests are most severe from about June until late August.

The best method of control is to spray when the flowers first open and repeat about two weeks later. Use a preparation such as sevin.

Peas are subject to attacks by the pea and bean weevil (top), the pea moth (center) and its grub (bottom).

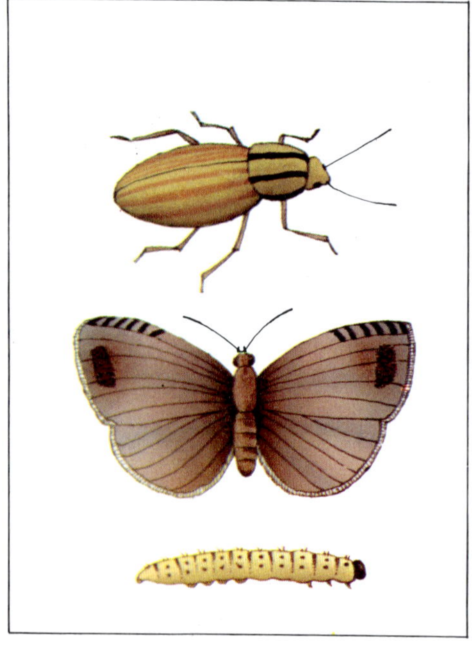

Pest damage symptoms: Weevils eat edges of leaves while the pea moth grub attacks the peas in their pods.

Dusting with sevin will control the pea and bean weevil, to obtain a good crop such as this.

Pea and bean weevils Although a scalloped edge to leaves may look quite attractive, it is, unfortunately, a sign that your plants are being infested by the pea or bean weevil, which devours leaf edges in this fashion.

March to June are the months when attacks are most severe. To control them, dust with sevin. It is mostly the younger plants which are affected.

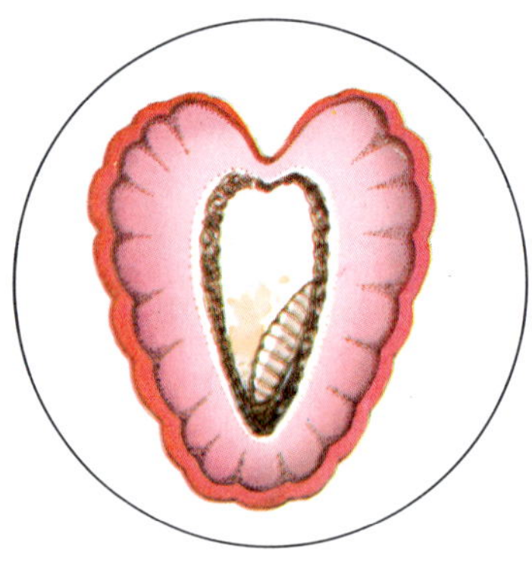

Small white grubs in the fruit of raspberries are the sign of a raspberry beetle infestation.

Raspberry beetle Little white grubs feeding in raspberry fruits are a sign of attack by the raspberry beetle. Fruits can be quickly ruined unless precautions are taken.

June to August are the months when attacks are most likely to occur. Use malathion to control. The best time to start spraying is as soon as the early fruits begin to color.

Without precautions, healthy raspberries like the above could swiftly be ruined by an attack of raspberry beetle.

Slugs (top) and snails (bottom) are usually found in poorly drained soils and weedy gardens.

Slugs and snails Many plants are affected by these pests, particularly tender ones such as lettuce and strawberries. Irregular holes are eaten into the leaves and there are always tell-tale slime tracks too.

Slugs and snails are very troublesome where the ground is badly drained and wet and where there are weeds and tall grass nearby. Control is best achieved by the use of slug pellets or liquid. However, make sure to keep pets away from baited areas.

Wireworms This is a particularly troublesome pest, especially in new gardens where Wireworms were previously infesting the grass.

The tough-skinned larvae, which are about $1/2$-$3/4$ in. long and yellowish in color, are to be found in the fleshy roots of crops such as carrots and potatoes, where they tunnel away until the plant wilts and dies.

As the garden is cultivated, especially in new sites, work in quantities of rotenone dust.

Wireworms are troublesome potato pests that eat their way into the tubers. They also attack other root crops.

The grey or brown armour-plated woodlouse is usually only troublesome in greenhouses or garden frames.

Woodlouse These pests feed at night and are found in damp, dark hiding places under stones, loose bark and flower pots. They are only a serious threat to plants growing under glass, where they feed on the roots, leaves and stems. To control them, eliminate their hiding places and dust with Sevin.

Woolly aphis If tufts of a near-white wool-ly substance are noticed on the stems and branches of plants, it is a sure sign of woolly

aphis attack. Apples and several other top-fruits are affected as well as some shrubs. These attacks usually occur from April to late September.

The best method of control is to spray a broad spectrum insecticide. If you are spraying apple or pear trees, make certain the chemical can be used on food crops.

Woolly aphis attack is identifiable by the tufts of white woolly substance that form on the stems of plants.

Index

aphids, 47, 55
apple, 9, 27, 28, 29, 31, 49, 56, 58, 63
apple sawfly, 56

basic slag, 10–11
beans, 12, 21, 50, 60
bean weevil, 60
beetroot, 12
big bud, 56
bindweed, 44
black spot, 49
blackberry, 36
blackcurrant, 32, 56
blight, 49
bonemeal, 10–11
botrytis, 51
broccoli, 12
brussels sprouts, 7, 12, 50, 56
buds, 28

cabbage, 7, 10, 12, 50, 51, 56
cabbage root fly, 56
calcium, 6, 8, 9
canker, 49
capillary mat, 25
capsid bug, 56–7
carrot, 9, 12, 21, 57, 62
carrot fly, 57
caterpillar, 57
cauliflower, 50, 56
chalk, 10–11
cherry, 33
chickweed, 42
chocolate spot, 50
chrysanthemum, 51, 56, 58, 59
cinquefoil, 42
clematis, 38, 39
clover, 42
club root, 50
codling moth, 58
compost, 4, 16–21
    accelerator, 18–19
    containers, 18
    heap, construction of, 18–19
    John Innes, 13, 51
    materials for, 16–17
    used as mulch, 21
couch grass, 44
creeping buttercup, 42
cucumber, 21, 22, 55
cutworms, 58

dahlia, 56, 59
damping off, 47, 51
damson, 37–8
dandelion, 42
diseases and disease control, 47–54
dock, 44

earwig, 59

fertilizers, 5, 7, 9–15
    application of, 14–15
    list of, 10–11

mixtures, 12–13
    organic, 6, 14
field woodrush, 43
foliar feeding, 14
fruit trees, 9, 12
    pruning, 26–38

gooseberry, 33, 56
grapes, 36
greenhouse,
    compost for, 21
    watering, 25
grey mould, 51
ground elder, 45
groundsel, 45
gypsum, 10–11

heather, 7
hollyhock, 52
horsetail, 45
humus, *see* compost
hydrangea, 7

lawn,
    compost for, 21
    fertilizers for, 13, 14–15
    watering, 24
    weeds in, 42–3
leaf miner, 59
leaf mould, 52
leatherjackets, 59
leek, 60
lettuce, 10, 20, 51, 62
lime, 10–11
loganberry, 36

manure, 14
meadowgrass, 43
melon, 22, 55
mildew, 47
mint, 52
moisture indicator, 23
moth,
    codling, 58
    pea, 60
mulching, 21, 24

nectarine, 34–5
nettle, 45
nitrates, 10–11, 12
nitro-chalk, 10–11
nitrogen, 6, 8, 9
nutrients, *see* plant foods

onion, 60
onion fly, 60
oxalis, 45

parsnip, 12, 49
pea moth, 60
pea weevil, 60
peach, 34–5
pear, 9, 29, 31, 58
pearlwort, 43
peas, 12, 21, 60
pests and pest control, 5, 47, 48,
    55–63

petty spurge, 46
phosphorus, 6, 8, 9
plant foods, 6–15
    basic chemicals, 6, 7, 8, 9
    fertilizers, 9–15
    list of, 10–11
plantain, 46
plum, 37–8
potash, 9
potassium, 6
potato, 9, 49, 52, 53, 62
pruning, 26–40

raspberry, 26, 33, 62
raspberry beetle, 62
redcurrant, 32, 56
rhododendron, 7
ribwort, 43
root crops, 9, 12, 21
rose, 12, 39–40, 49, 52
rust, 52

saltpetre, 12
scab, 52
seaweed, 14
shallot, 60
shepherd's purse, 46
shrubs, 38–9, 63
slug, 62
snail, 62
soil,
    acidity, 7, 9
    nutrients, testing for, 7
speedwell, 43
spot treatment, 44
spraying, *see* pest control
strawberry, 62
sulphates, 10–11, 12
superphosphate of lime, 10–11

thistle, 46
tomato, 12, 21, 22, 49, 51, 52, 55
top dressing, 21
trace elements, 6, 14
trees, 5, 30, 31, 40, *see also*
    fruit trees
trefoil, 43
turnip, 50

verticillium wilt, 53
vine, 36–7
virus, 54

wallflower, 50
watering, 22–5
weeds and weedkillers, 4, 41–6
    in gardens, 44–6
    in lawns, 42–3
    safety, 48
weevils, 60
white currant, 32, 56
wireworms, 62
wisteria, 38, 39
woodlouse, 63
woolly aphis, 63

yarrow, 43